Fundamentals of Dynamics and Control of Space Systems

Solution Manual

Krishna Dev Kumar

ii

Dr. Krishna. D. Kumar
Professor and Canada Research Chair in Space Systems
Department of Aerospace Engineering
Ryerson University
Toronto, Ontario
Canada M5B 2K3
Email: kdkumar@ryerson.ca
http://www.ryerson.ca/ kdkumar

Cover page: Full view of the International Space Station as photographed from the Space Shuttle Discovery during the STS-114 Return to Flight mission, following the undocking of the two spacecraft. (Courtesy of NASA)

Fundamentals of Dynamics and Control of Space Systems

Solution Manual

Krishna Dev Kumar
Professor and Canada Research Chair in Space Systems
Department of Aerospace Engineering
Ryerson University
Toronto, Canada

iv

Preface

This Solution Manual is prepared to accompany and supplement the author's text "Fundamentals of Dynamics and Control of Space Systems" by Krishna Dev Kumar, 2012. It contains detailed solutions for most problems in the textbook.

September 3, 2012   Krishna Kumar

Contents

Chapter 2

Kinematics, Momentum and Energy

Problem Set 2

2.1 The coordinate frames used in studying the dynamics of a spacecraft are as follows:

a) Inertial reference frame,

b) Orbital reference frame,

c) Perifocal reference frame,

c) Satellite body-fixed reference frame.

2.2 The inertial frames are those coordinate frames that are nonrotating and nonaccelerating frames. The inertial frames are relevant because in applying the Newton's second law of motion

$$\vec{F} = m\frac{d\vec{V}}{dt} \tag{2.1}$$

to derive the equation of motion of a system, the velocity $\vec{V}$ and the corresponding acceleration $d\vec{V}/dt$ in the right-hand side of the above equation are to measured with respect to an inertial frame of reference.

An Earth-fixed frame is not an inertial frame as it is spinning about its axis with a period of 24 hour. When viewed from space, the point on the surface of the earth moves in a circle as the earth spins on its axis. Thus, it is accelerating with an centripetal acceleration of $r\omega^2$, where r is the position of the point of the Earth center of mass and ω is the rate of spin of the Earth. With the earth a point on its surface also orbits the Sun. With the solar system, it orbits the center of the galaxy. Thus, the Earth-fixed frame is an accelerating frame and not an inertial frame.

We consider just the effect of the spinning motion of the Earth and therefore the inertial acceleration can be written as

$$\left.\frac{d\vec{V}}{dt}\right|_{inertial} = \left.\frac{d\vec{V}}{dt}\right|_{body} + \vec{\omega} \times \vec{V}_{body} \tag{2.2}$$

The corresponding error in considering an Earth-fixed frame as an inertial frame is

$$\text{Error} = \left.\frac{d\vec{V}}{dt}\right|_{inertial} - \left.\frac{d\vec{V}}{dt}\right|_{body} = \vec{\omega} \times \vec{V}_{body} \tag{2.3}$$

The Earth's spin rate ω is

$$\vec{\omega} = \omega_k \hat{k} = -\frac{2\pi}{T}\hat{k}$$
$$= -\frac{2\pi}{24 \times 3600}\hat{k} = 7.275 \times 10^{-5}\hat{k} \tag{2.4}$$

where $\hat{k}$ is a unit vector along the z-direction as taken for the aircraft body-fixed frame.

The order of magnitude error would be $10^{-4} \times V_{body}$. As this magnitude is usually very small when compared to the magnitude of other relevant accelerations like the gravitational acceleration, which is 9.81 m/s^2, and we often treat the Earth-fixed frame as an inertial frame. when solving problems.

2.3 The inertial position vectors for spacecraft m_1 and m_2 are

$$\vec{R}_1 = \vec{R} - \gamma\vec{L} \tag{2.5}$$
$$\vec{R}_2 = \vec{R} + (1 - \gamma)\vec{L} \tag{2.6}$$

where $\gamma = m_2/(m_1 + m_2)$. The corresponding magnitudes are

$$R_1 = [R^2 + \gamma^2 L^2 - 2\gamma\vec{R} \cdot \vec{L}]^{1/2} \tag{2.7}$$
$$R_2 = [R^2 + (1 - \gamma)^2 L^2 + 2(1 - \gamma)\vec{R} \cdot \vec{L}]^{1/2} \tag{2.8}$$

where $L = L_0 + vt$. The nomenclature L_o defines the initial length of the cable while v is the speed by which the length of the cable varies.

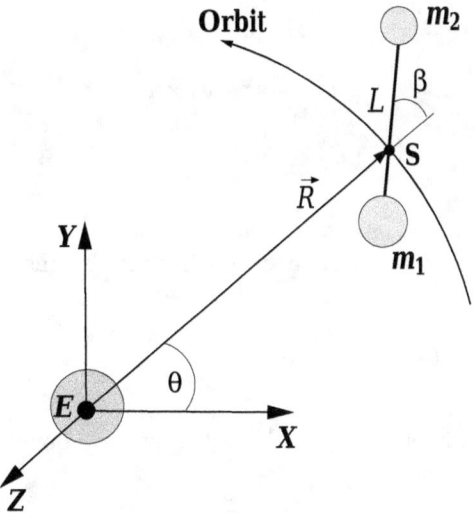

Figure 2.1: A dumbbell satellite system undergoing in-plane libration.

Expressing $\vec{R}$ and $\vec{L}$ in terms of unit vectors of the respective coordinate frames as

$$\vec{R} = R\hat{i}_o, \quad \vec{L} = L\hat{i} \qquad (2.9)$$

Applying the transformation between coordinate frames $S-x_oy_oz_o$ and $S-xyz$, we get

$$\hat{i}_o \cdot \hat{i} = cos\beta \qquad (2.10)$$

and using this relation in Eqs. (2.7) and (2.8), we have the inertial positions of the spacecraft m_1 and m_2 as

$$R_1 = [R^2 + \gamma^2 L^2 - 2\gamma RL cos\beta]^{1/2} \qquad (2.11)$$

$$R_2 = [R^2 + (1-\gamma)^2 L^2 + 2(1-\gamma)RL cos\beta]^{1/2} \qquad (2.12)$$

The inertial velocity vectors for spacecraft m_1 and m_2 are

$$\vec{V}_1 = \dot{\vec{R}}_1 = \dot{\vec{R}} - \gamma\dot{\vec{L}} \qquad (2.13)$$

$$\vec{V}_2 = \dot{\vec{R}}_2 = \dot{\vec{R}} + (1-\gamma)\dot{\vec{L}} \qquad (2.14)$$

The corresponding magnitudes are

$$V_1 = [\dot{\vec{R}}^2 + \gamma^2 \dot{\vec{L}}^2 - 2\gamma\dot{\vec{R}} \cdot \dot{\vec{L}}]^{1/2} \qquad (2.15)$$

$$V_2 = [\dot{\vec{R}}^2 + (1-\gamma)^2 \dot{\vec{L}}^2 + 2(1-\gamma)\dot{\vec{R}} \cdot \dot{\vec{L}}]^{1/2} \qquad (2.16)$$

$\dot{\vec{R}}$ and $\dot{\vec{L}}$ can be written as

$$\dot{\vec{R}} = \left(\dot{\vec{R}}\right)_{x_o y_o z_o} + \vec{\omega}_o \times \vec{R} \qquad (2.17)$$

$$\dot{\vec{L}} = \left(\dot{\vec{L}}\right)_{xyz} + \vec{\omega} \times \vec{L} \qquad (2.18)$$

Knowing the system is orbiting in a circular orbit (*i.e.*, $\dot{R} = 0$), and the cable connecting the two spacecraft is moving with a constant speed of v, we get

$$\left(\dot{\vec{R}}\right)_{x_o y_o z_o} = 0, \quad \left(\dot{\vec{L}}\right)_{xyz} = \vec{v} \qquad (2.19)$$

where $\vec{v}$ is in the direction of $\vec{L}$. Substituting the above relations in Eqs. (2.17)-(2.18), we obtain

$$\dot{\vec{R}} = \vec{\omega}_o \times \vec{R} \qquad (2.20)$$

$$\dot{\vec{L}} = \vec{v} + \vec{\omega} \times \vec{L} \qquad (2.21)$$

The terms $\dot{\vec{R}}^2$ and $\dot{\vec{L}}^2$ can be written as

$$\dot{\vec{R}}^2 = (\vec{\omega}_o \times \vec{R})^2 \qquad (2.22)$$

$$\dot{\vec{L}}^2 = v^2 + (\vec{\omega} \times \vec{L})^2 + 2\vec{v} \cdot (\vec{\omega} \times \vec{L}) \qquad (2.23)$$

Writing $\vec{\omega}_o$, $\vec{R}$, $\vec{\omega}$, $\vec{L}$, and $\vec{v}$ in terms of the unit vectors of the respective coordinate frames, we have

$$\omega_o = \dot{\theta}\hat{k}_o, \quad \vec{R} = R\hat{i}_o, \quad \vec{\omega} = (\dot{\theta} + \dot{\beta})\hat{k}, \quad \vec{L} = L\hat{i}, \quad \vec{v} = v\hat{i} \qquad (2.24)$$

Inserting these expressions into Eqs. (2.22)-(2.23) and solving, we have

$$\dot{\vec{R}}^2 = \dot{\theta}^2 R^2 \qquad (2.25)$$

$$\dot{\vec{L}}^2 = v^2 + (\dot{\theta} + \dot{\beta})^2 L^2 \qquad (2.26)$$

Note that as $\vec{v} \perp (\vec{\omega} \times \vec{L})$, $\vec{v} \cdot (\vec{\omega} \times \vec{L}) = 0$.

Next we derive $\dot{\vec{R}} \cdot \dot{\vec{L}}$. Using Eqs. (2.20)-(2.21), we can write

$$\begin{aligned} \dot{\vec{R}} \cdot \dot{\vec{L}} &= (\vec{\omega}_o \times \vec{R}) \cdot (\vec{v} + \vec{\omega} \times \vec{L}) \\ &= (\vec{\omega}_o \times \vec{R}) \cdot \vec{v} + (\vec{\omega}_o \times \vec{R}) \cdot (\vec{\omega} \times \vec{L}) \\ &= \dot{\theta} R v(\hat{j}_o \cdot \hat{i}) + \dot{\theta}(\dot{\theta} + \dot{\beta}) R L(\hat{j}_o \cdot \hat{j}) \end{aligned} \qquad (2.27)$$

From the coordinate transformation between the coordinate frame $S - \hat{i}_o\hat{j}_o\hat{k}_o$ and $S - \hat{i}\hat{j}\hat{k}$, we have

$$\hat{j}_o \cdot \hat{i} = sin\beta \tag{2.28}$$

$$\hat{j}_o \cdot \hat{j} = cos\beta \tag{2.29}$$

Thus, we can express $\dot{\vec{R}} \cdot \dot{\vec{L}}$ as

$$\dot{\vec{R}} \cdot \dot{\vec{L}} = \dot{\theta}Rvsin\beta + \dot{\theta}(\dot{\theta} + \dot{\beta})RLcos\beta \tag{2.30}$$

Substituting the expressions for $\dot{\vec{R}}^2$ and $\dot{\vec{L}}^2$ from Eqs. (2.25)-(2.26) and the expression for $\dot{\vec{R}} \cdot \dot{\vec{L}}$ from Eq. (2.30) into Eqs. (2.15)-(2.16), we finally obtain the magnitudes of the inertial velocity vectors for spacecraft m_1 and m_2 as

$$V_1 = \{\dot{\theta}^2 R^2 + \gamma^2(\dot{L}^2 + (\dot{\theta} + \dot{\beta})^2 L^2) - 2\gamma\dot{\theta}R[vsin\beta + (\dot{\theta} + \dot{\beta})Lcos\beta]\}^{1/2} \tag{2.31}$$

$$V_2 = \{\dot{\theta}^2 R^2 + (1-\gamma)^2(\dot{L}^2 + (\dot{\theta} + \dot{\beta})^2 L^2)$$
$$+ 2(1-\gamma)\dot{\theta}R[vsin\beta + (\dot{\theta} + \dot{\beta})Lcos\beta]\}^{1/2} \tag{2.32}$$

The inertial acceleration vectors for the spacecraft m_1 and m_2 are written using Eqs. (2.13)-(2.14) for their velocity vectors, as

$$\vec{a}_1 = \dot{\vec{V}}_1 = \ddot{\vec{R}} - \gamma\ddot{\vec{L}} \tag{2.33}$$

$$\vec{a}_2 = \dot{\vec{V}}_2 = \ddot{\vec{R}} + (1-\gamma)\ddot{\vec{L}} \tag{2.34}$$

Here $\ddot{\vec{R}}$ and $\ddot{\vec{L}}$ can be expressed as

$$\ddot{\vec{R}} = \ddot{\vec{R}}_{x_oy_oz_o} + 2(\vec{\omega}_o \times \dot{\vec{R}}_{x_oy_oz_o}) + \vec{\omega}_o \times (\vec{\omega}_o \times \vec{R}) + \dot{\vec{\omega}}_o \times \vec{R} \tag{2.35}$$

$$\ddot{\vec{L}} = \ddot{\vec{L}}_{xyz} + 2(\vec{\omega} \times \dot{\vec{L}}_{xyz}) + \vec{\omega} \times (\vec{\omega} \times \vec{L}) + \dot{\vec{\omega}} \times \vec{L} \tag{2.36}$$

The centripetal acceleration components $\vec{\omega}_o \times (\vec{\omega}_o \times \vec{R})$ and $\vec{\omega} \times (\vec{\omega} \times \vec{L})$ can be simplified using the triple vector product relation $\vec{a} \times (\vec{b} \times \vec{c}) = (\vec{a} \cdot \vec{c})\vec{b} - (\vec{a} \cdot \vec{b})\vec{c}$, as

$$\vec{\omega}_o \times (\vec{\omega}_o \times \vec{R}) = (\vec{\omega}_o \cdot \vec{R})\vec{\omega}_o - \omega_o^2\vec{R} = -\omega_o^2\vec{R} \tag{2.37}$$

$$\vec{\omega}_o \times (\vec{\omega}_o \times \vec{L}) = (\vec{\omega}_o \cdot \vec{L})\vec{\omega}_o - \omega_o^2\vec{L} = -\omega^2\vec{L} \tag{2.38}$$

Note the terms $\vec{\omega}_o \cdot \vec{R} = 0$ and $(\vec{\omega}_o \cdot \vec{L}) = 0$ since $\omega_o \perp \vec{R}$ and $\omega \perp \vec{L}$.

Knowing that the system is in a circular orbit and the cable connecting the two spacecraft is deployed with constant velocity, i.e.,

$$\dot{\vec{R}}_{x_oy_oz_o} = \ddot{\vec{R}}_{x_oy_oz_o} = 0, \quad \dot{\vec{\omega}}_o = 0, \quad \ddot{\vec{L}}_{xyz} = \dot{v} = 0, \quad \dot{\vec{\omega}} = \ddot{\beta}\hat{k} \tag{2.39}$$

and writing all vectors in Eqs. (2.35)-(2.36) in terms of the unit vectors along the respective coordinate frames, we obtain

$$\ddot{\vec{R}} = -\dot{\theta}^2 R \hat{i}_o \tag{2.40}$$

$$
\begin{aligned}
\ddot{\vec{L}} &= 2(\dot{\theta} + \dot{\beta})v(\hat{k} \times \hat{i}) - (\dot{\theta} + \dot{\beta})^2 L \hat{i} + \ddot{\beta} L(\hat{k} \times \vec{i}) \\
&= 2(\dot{\theta} + \dot{\beta})v \hat{j} - (\dot{\theta} + \dot{\beta})^2 L \hat{i} + \ddot{\beta} L \hat{j} \\
&= -(\dot{\theta} + \dot{\beta})^2 L \hat{i} + [2(\dot{\theta} + \dot{\beta})v + \ddot{\beta} L] \hat{j} \tag{2.41}
\end{aligned}
$$

Substituting the above relations in Eqs. (2.33)-(2.34), we have the inertial accelerations as

$$\vec{a}_1 = -\dot{\theta}^2 R \hat{i}_o + \gamma(\dot{\theta} + \dot{\beta})^2 L \hat{i} - \gamma[2(\dot{\theta} + \dot{\beta})v + \ddot{\beta} L] \hat{j}\} \tag{2.42}$$

$$\vec{a}_2 = -\dot{\theta}^2 R \hat{i}_o - (1 - \gamma)(\dot{\theta} + \dot{\beta})^2 L \hat{i} - (1 - \gamma)[2(\dot{\theta} + \dot{\beta})v - \ddot{\beta} L] \hat{j}\} \tag{2.43}$$

The corresponding magnitude for spacecraft m_1 can be expressed using the algebraic relations $((a + b + c)^2 = a^2 + b^2 + c^2 + 2ab + 2ac + 2bc))$ as

$$
\begin{aligned}
a_1 = \Big\{ &(\dot{\theta}^2 R)^2 + \gamma^2(\dot{\theta} + \dot{\beta})^4 L^2 + \gamma^2[2(\dot{\theta} + \dot{\beta})v + \ddot{\beta} L]^2 \\
&- 2\gamma\dot{\theta}^2(\dot{\theta} + \dot{\beta})^2 RL(\hat{i}_o \cdot \hat{i}) + 2\dot{\theta}^2\gamma[2(\dot{\theta} + \dot{\beta})v + \ddot{\beta} L]R(\hat{i}_o \cdot \hat{j}) \Big\}^{1/2}
\end{aligned}
\tag{2.44}
$$

From the coordinate transformation between the coordinate frames $S - \hat{i}_o \hat{j}_o \hat{k}_o$ and $S - \hat{i}\hat{j}\hat{k}$, we get

$$\hat{i}_o \cdot \hat{i} = cos\beta, \quad \hat{i}_o \cdot \hat{j} = cos(90 + \beta) = -sin\beta \tag{2.45}$$

Using these relations, the corresponding magnitude for spacecraft m_1 can be expressed as

$$
\begin{aligned}
a_1 = \Big\{ &(\dot{\theta}^2 R)^2 + \gamma^2(\dot{\theta} + \dot{\beta})^4 L^2 + \gamma^2[2(\dot{\theta} + \dot{\beta})v + \ddot{\beta} L]^2 \\
&- 2\gamma\dot{\theta}^2(\dot{\theta} + \dot{\beta})^2 RLcos\beta - 2\dot{\theta}^2\gamma[2(\dot{\theta} + \dot{\beta})v + \ddot{\beta} L]Rsin\beta \Big\}^{1/2}
\end{aligned}
\tag{2.46}
$$

Similarly, the acceleration of spacecraft m_2 can be obtained as

$$
\begin{aligned}
a_2 = \Big\{ &(\dot{\theta}^2 R)^2 + (1 - \gamma)^2(\dot{\theta} + \dot{\beta})^4 L^2 + (1 - \gamma)^2[2(\dot{\theta} + \dot{\beta})v - \ddot{\beta} L]^2 \\
&+ 2(1 - \gamma)\dot{\theta}^2(\dot{\theta} + \dot{\beta})^2 RLcos\beta - 2\dot{\theta}^2(1 - \gamma)[2(\dot{\theta} + \dot{\beta})v - \ddot{\beta} L]Rsin\beta \Big\}^{1/2}
\end{aligned}
\tag{2.47}
$$

Note the preceding expression can be derived simply by replacing replacing γ by $(1 - \gamma)$ and L by $-L$ in Eq. (2.46).

2.4 The inertial position vectors for spacecraft m_1 and m_2 are

$$\vec{R}_1 = \vec{R} - \gamma \vec{L} \qquad (2.48)$$

$$\vec{R}_2 = \vec{R} + (1-\gamma)\vec{L} \qquad (2.49)$$

where $\gamma = m_2/(m_1 + m_2)$. The corresponding magnitudes are

$$R_1 = [R^2 + \gamma^2 L^2 - 2\gamma \vec{R} \cdot \vec{L}]^{1/2} \qquad (2.50)$$

$$R_2 = [R^2 + (1-\gamma)^2 L^2 + 2(1-\gamma)\vec{R} \cdot \vec{L}]^{1/2} \qquad (2.51)$$

Writing $\vec{R}$ and $\vec{L}$ in terms of the unit vectors of the respective coordinate frames, we have

$$\vec{R} = R\hat{i}_o; \quad \vec{L} = L\hat{i} \qquad (2.52)$$

To express $\hat{i}$ in terms of unit vectors in the frame $S - x_o y_o z_o$, we consider the transformation as

$$\left\{ \begin{array}{c} \hat{i} \\ \hat{j} \\ \hat{k} \end{array} \right\} = R_{zy}(\beta, \eta) \left\{ \begin{array}{c} \hat{i}_o \\ \hat{j}_o \\ \hat{k}_o \end{array} \right\} \qquad (2.53)$$

where $R_{zy}(\beta, \eta)$ is

$$R_{zy}(\beta, \eta) = R_y(\eta) R_z(\beta) = \begin{bmatrix} cos\eta & 0 & -sin\eta \\ 0 & 1 & 0 \\ sin\eta & 0 & cos\eta \end{bmatrix} \begin{bmatrix} cos\beta & sin\beta & 0 \\ -sin\beta & cos\beta & 0 \\ 0 & 0 & 1 \end{bmatrix}$$

$$= \begin{bmatrix} cos\beta cos\eta & sin\beta cos\eta & -sin\eta \\ -sin\beta & cos\beta & 0 \\ cos\beta sin\eta & sin\beta sin\eta & cos\eta \end{bmatrix} \qquad (2.54)$$

Using Eq. (2.52), we can write $\hat{i}$ as

$$\hat{i} = cos\beta cos\eta \hat{i}_o + sin\beta cos\eta \hat{j}_o - sin\eta \hat{k}_o \qquad (2.55)$$

Applying the above relation and using Eqs. (2.52), we obtain the magnitudes of the position vectors as

$$R_1 = [R^2 + \gamma^2 L^2 - 2\gamma RL cos\beta cos\eta]^{1/2} \qquad (2.56)$$

$$R_2 = [R^2 + (1-\gamma)^2 L^2 + 2(1-\gamma)RL cos\beta cos\eta]^{1/2} \qquad (2.57)$$

The inertial velocity vectors for spacecraft m_1 and m_2 are

$$\vec{V}_1 = \dot{\vec{R}}_1 = \dot{\vec{R}} - \gamma \dot{\vec{L}} \qquad (2.58)$$

$$\vec{V}_2 = \dot{\vec{R}}_2 = \dot{\vec{R}} + (1 - \gamma)\dot{\vec{L}} \qquad (2.59)$$

The corresponding magnitudes are

$$V_1 = [\dot{\vec{R}}^2 + \gamma^2 \dot{\vec{L}}^2 - 2\gamma \dot{\vec{R}} \cdot \dot{\vec{L}}]^{1/2} \qquad (2.60)$$

$$V_2 = [\dot{\vec{R}}^2 + (1 - \gamma)^2 \dot{\vec{L}}^2 + 2(1 - \gamma)\dot{\vec{R}} \cdot \dot{\vec{L}}]^{1/2} \qquad (2.61)$$

The $\dot{\vec{R}}$ and $\dot{\vec{L}}$ can be written as

$$\dot{\vec{R}} = \left(\dot{\vec{R}}\right)_{x_o y_o z_o} + \vec{\omega}_o \times \vec{R} \qquad (2.62)$$

$$\dot{\vec{L}} = \left(\dot{\vec{L}}\right)_{xyz} + \vec{\omega} \times \vec{L} \qquad (2.63)$$

Knowing the system is orbiting in a circular orbit (*i.e.*, $\dot{R} = 0$), and length of the cable connecting the two spacecraft is constant, we get

$$\left(\dot{\vec{R}}\right)_{x_o y_o z_o} = 0, \quad \left(\dot{\vec{L}}\right)_{xyz} = 0 \qquad (2.64)$$

Substituting the above relations in Eqs. (2.62)-(2.63), we obtain

$$\dot{\vec{R}} = \vec{\omega}_o \times \vec{R} \qquad (2.65)$$

$$\dot{\vec{L}} = \vec{\omega} \times \vec{L} \qquad (2.66)$$

The terms $\dot{\vec{R}}^2$ and $\dot{\vec{L}}^2$ can be written as

$$\dot{\vec{R}}^2 = (\vec{\omega}_o \times \vec{R})^2 \qquad (2.67)$$

$$\dot{\vec{L}}^2 = (\vec{\omega} \times \vec{L})^2 \qquad (2.68)$$

Writing $\vec{\omega}_o$, $\vec{R}$, $\vec{\omega}$, $\vec{L}$, and $\vec{v}$ in terms of the unit vectors of the respective coordinate frames, we have

$$\omega_o = \dot{\theta}\hat{k}_o, \quad \vec{R} \;\; = R\hat{i}_o, \quad \vec{\omega} = (\dot{\theta} + \dot{\beta})\hat{k}_o + \dot{\eta}\hat{j}, \quad \vec{L} = L\hat{i} \qquad (2.69)$$

From the coordinate transformation between the coordinate frame $S - \hat{i}_o\hat{j}_o\hat{k}_o$ and $S - \hat{i}\hat{j}\hat{k}$, we have

$$\hat{k}_o = -sin\eta\hat{i} + cos\eta\hat{k} \qquad (2.70)$$

Using the preceding equation, we write $\vec{\omega}$ in terms of unit vectors along the rotating coordinate frame $S - xyz$ as

$$\vec{\omega} = \omega_x \hat{i} + \omega_y \hat{j} + \omega_z \hat{k} \tag{2.71}$$

where

$$\omega_x = -(\dot{\theta} + \dot{\beta})sin\eta, \quad \omega_y = \dot{\eta}, \quad \omega_z = (\dot{\theta} + \dot{\beta})cos\eta \tag{2.72}$$

Inserting the expressions given by Eqs. (2.69) and (2.71) into Eqs. (2.67)-(2.68) and solving, we have

$$\dot{\vec{R}}^2 = \dot{\theta}^2 R^2 \tag{2.73}$$

$$\dot{\vec{L}}^2 = [(\dot{\theta} + \dot{\beta})^2 + \dot{\eta}^2]L^2 \tag{2.74}$$

Next we derive $\dot{\vec{R}} \cdot \dot{\vec{L}}$. Using Eqs. (2.65)-(2.66), we can write

$$\begin{aligned}
\dot{\vec{R}} \cdot \dot{\vec{L}} &= (\vec{\omega}_o \times \vec{R}) \cdot (\vec{\omega} \times \vec{L}) \\
&= \dot{\theta} RL\{\hat{j}_o \cdot [-\dot{\eta}\hat{k} + (\dot{\theta} + \dot{\beta})cos\eta\hat{j}]\} \\
&= \dot{\theta} RL[-\dot{\eta}(\hat{j}_o \cdot \hat{k}) + (\dot{\theta} + \dot{\beta})cos\eta(\hat{j}_o \cdot \hat{j})]
\end{aligned} \tag{2.75}$$

From the coordinate transformation between the coordinate frame $S - \hat{i}_o\hat{j}_o\hat{k}_o$ and $S - \hat{i}\hat{j}\hat{k}$, we have

$$\hat{j}_o \cdot \hat{j} = cos\beta, \quad \hat{j}_o \cdot \hat{k} = sin\beta sin\eta \tag{2.76}$$

Thus, we can express $\dot{\vec{R}} \cdot \dot{\vec{L}}$ as

$$\dot{\vec{R}} \cdot \dot{\vec{L}} = \dot{\theta} RL[-\dot{\eta}sin\beta sin\eta + (\dot{\theta} + \dot{\beta})cos\beta cos\eta] \tag{2.77}$$

Substituting the expressions for $\dot{\vec{R}}^2$, $\dot{\vec{L}}^2$ from Eqs. (2.73)-(2.74) and the expression for $\dot{\vec{R}} \cdot \dot{\vec{L}}$ from Eq. (2.77) into Eqs. (2.60)-(2.61), we finally obtain the magnitudes of the inertial velocity vectors for spacecraft m_1 and m_2 as

$$\begin{aligned}
V_1 = &\{\dot{\theta}^2 R^2 + \gamma^2\{\dot{L}^2 + [(\dot{\theta} + \dot{\beta})^2 + \dot{\eta}^2]L^2\} \\
&- 2\gamma\dot{\theta} RL[-\dot{\eta}sin\beta sin\eta + (\dot{\theta} + \dot{\beta})cos\beta cos\eta]\}^{1/2}
\end{aligned} \tag{2.78}$$

$$\begin{aligned}
V_2 = &\{\dot{\theta}^2 R^2 + (1 - \gamma)^2\{\dot{L}^2 + [(\dot{\theta} + \dot{\beta})^2 + \dot{\eta}^2]L^2\} \\
&+ 2(1 - \gamma)\dot{\theta} RL[-\dot{\eta}sin\beta sin\eta + (\dot{\theta} + \dot{\beta})cos\beta cos\eta]\}^{1/2}
\end{aligned} \tag{2.79}$$

The inertial acceleration vectors for the spacecraft m_1 and m_2 are written using Eqs. (2.58)-(2.59) for their velocity vectors, as

$$\vec{a}_1 = \dot{\vec{V}}_1 = \ddot{\vec{R}} - \gamma \ddot{\vec{L}} \tag{2.80}$$

$$\vec{a}_2 = \dot{\vec{V}}_2 = \ddot{\vec{R}} + (1 - \gamma) \ddot{\vec{L}} \tag{2.81}$$

Here $\ddot{\vec{R}}$ and $\ddot{\vec{L}}$ can be expressed as

$$\ddot{\vec{R}} = \ddot{\vec{R}}_{x_o y_o z_o} + 2(\vec{\omega}_o \times \dot{\vec{R}}_{x_o y_o z_o}) + \vec{\omega}_o \times (\vec{\omega} \times \vec{R}) + \dot{\vec{\omega}}_o \times \vec{R} \tag{2.82}$$

$$\ddot{\vec{L}} = \ddot{\vec{L}}_{xyz} + 2(\vec{\omega} \times \dot{\vec{L}}_{xyz}) + \vec{\omega} \times (\vec{\omega} \times \vec{L}) + \dot{\vec{\omega}} \times \vec{L} \tag{2.83}$$

Knowing that the system is in a circular orbit and the cable connecting the two spacecraft are moving with constant velocity, *i.e.*,

$$\dot{\vec{R}}_{x_o y_o z_o} = \ddot{\vec{R}}_{x_o y_o z_o} = 0, \quad \dot{\vec{\omega}}_o = 0, \quad \dot{\vec{L}}_{xyz} = \ddot{\vec{L}}_{xyz} = 0 \tag{2.84}$$

and writing all vectors in terms of the unit vectors along the respective coordinate frames, we obtain

$$\ddot{\vec{R}} = -\dot{\theta}^2 R \hat{i}_o \tag{2.85}$$

$$\tag{2.86}$$

and derive the following terms for $\ddot{\vec{L}}$ using $\vec{\omega}$ (given by Eq. (2.71)):

$$\vec{\omega} \times (\vec{\omega} \times \vec{L}) = (\vec{\omega} \cdot \vec{L})\vec{\omega} - \omega^2 \vec{L} = \omega_x L \vec{\omega} - \omega^2 \vec{L} \tag{2.87}$$

$$\dot{\vec{\omega}} \times \vec{L} = [-\dot{\omega}_y \hat{k} + \dot{\omega}_z \hat{j}]L \tag{2.88}$$

where

$$\dot{\omega}_y = \ddot{\eta}, \quad \dot{\omega}_z = [\ddot{\beta} cos\eta - (\dot{\theta} + \dot{\beta})\dot{\eta} sin\eta] \tag{2.89}$$

So, the term $\ddot{\vec{L}}$ can be expressed as

$$\ddot{\vec{L}} = a_{L_x} \hat{i} + a_{L_y} \hat{j} + a_{L_z} \hat{k} \tag{2.90}$$

where

$$a_{L_x} = [\omega_x^2 - \omega^2]L, \quad a_{L_y} = [\omega_x \omega_y + \dot{\omega}_z]L, \quad a_{L_z} = [\omega_x \omega_z - \dot{\omega}_y]L \tag{2.91}$$

Substituting the above relations for $\ddot{\vec{R}}$ and $\ddot{\vec{L}}$ in Eqs. (2.80)-(2.81), we have

$$\vec{a}_1 = -\dot{\theta}^2 R \hat{i}_o - \gamma [a_{L_x}\hat{i} + a_{L_y}\hat{j} + a_{L_y}\hat{k}] \tag{2.92}$$

$$\vec{a}_2 = -\dot{\theta}^2 R \hat{i}_o + (1 - \gamma)[a_{L_x}\hat{i} + a_{L_y}\hat{j} + a_{L_y}\hat{k}]\} \tag{2.93}$$

The corresponding magnitude for spacecraft m_1 can be expressed as

$$a_1 = \Big\{ (\dot{\theta}^2 R)^2 + \gamma^2 [a_{L_x}^2 + a_{L_y}^2 + a_{L_z}^2]$$
$$- 2\dot{\theta}^2 R \gamma [a_{L_x}\hat{i}_o \cdot \hat{i} + a_{L_y}\hat{i}_o \cdot \hat{j} + a_{L_z}\hat{i}_o \cdot \hat{k}]\Big\}^{1/2} \tag{2.94}$$

From the coordinate transformation between the coordinate frame $S - \hat{i}_o\hat{j}_o\hat{k}_o$ and $S - \hat{i}\hat{j}\hat{k}$, we get

$$\hat{i}_o \cdot \hat{i} = cos\beta cos\eta, \quad \hat{i}_o \cdot \hat{j} = cos(90 + \beta) = -sin\beta, \quad \hat{i}_o \cdot \hat{k} = cos\beta sin\eta \tag{2.95}$$

Thus, the acceleration for spacecraft m_1 is

$$a_1 = \Big\{ (\dot{\theta}^2 R)^2 + \gamma^2 [a_{L_x}^2 + a_{L_y}^2 + a_{L_z}^2]$$
$$- 2\dot{\theta}^2 R \gamma [a_{L_x} \cos\beta cos\eta - a_{L_y} sin\beta + a_{L_z} cos\beta sin\eta]\Big\}^{1/2} \tag{2.96}$$

Similarly, the acceleration for spacecraft m_2 can be derived as

$$a_2 = \Big\{ (\dot{\theta}^2 R)^2 + (1 - \gamma)^2 [a_{L_x}^2 + a_{L_y}^2 + a_{L_z}^2]$$
$$+ 2\dot{\theta}^2 R(1 - \gamma)[a_{L_x} \cos\beta cos\eta - a_{L_y} sin\beta + a_{L_z} cos\beta sin\eta]\Big\}^{1/2} \tag{2.97}$$

2.5 The inertial position vectors of spacecraft m_1, m_2 and m_3 are

$$\vec{R}_1 = \vec{R} \tag{2.98}$$

$$\vec{R}_2 = \vec{R} + \vec{L}_1 \tag{2.99}$$

$$\vec{R}_3 = \vec{R}_2 + \vec{L}_2 \tag{2.100}$$

The corresponding magnitudes are

$$R_1 = R \tag{2.101}$$

$$R_2 = [R^2 + L_1^2 + 2\vec{R} \cdot \vec{L}_1]^{1/2} \tag{2.102}$$

$$R_3 = [R_2^2 + 2\vec{R} \cdot \vec{L}_2 + 2\vec{L}_1 \cdot \vec{L}_2]^{1/2} \tag{2.103}$$

Expressing $\vec{R}$, $\vec{L}_1$, and $\vec{L}_2$ in terms of unit vectors of their respective coordinate frames as

$$\vec{R} = R\hat{i}_o, \quad \vec{L}_1 = L_1\hat{i}_1, \quad \vec{L}_2 = L_2\hat{i}_2 \tag{2.104}$$

in the preceding equation and using the transformations between the coordinate frames $S-x_oy_oz_o$ and $S_1-x_1y_1z_1$, and the coordinate frames $S-x_oy_oz_o$ and $S_2-x_2y_2z_2$ as

$$\hat{i}_o \cdot \hat{i}_1 = cos\beta_1, \quad \hat{i}_o \cdot \hat{i}_2 = cos\beta_2, \quad \hat{i}_1 \cdot \hat{i}_2 = cos(\beta_2 - \beta_1) \qquad (2.105)$$

we obtain the positions of the spacecraft

$$R_1 = R \qquad (2.106)$$

$$R_2 = [R^2 + L_1^2 + 2RL_1cos\beta_1]^{1/2} \qquad (2.107)$$

$$R_3 = [R_2^2 + L_2^2 + 2RL_2cos\beta_2 + 2L_1L_2cos(\beta_2 - \beta_1)]^{1/2} \qquad (2.108)$$

The inertial velocity vectors for spacecraft m_1 and m_2 are

$$\vec{V}_1 = \dot{\vec{R}}_1 = \dot{\vec{R}} \qquad (2.109)$$

$$\vec{V}_2 = \dot{\vec{R}}_2 = \dot{\vec{R}} + \dot{\vec{L}}_1 \qquad (2.110)$$

$$\vec{V}_3 = \dot{\vec{R}}_3 = \vec{V}_2 + \dot{\vec{L}}_2 \qquad (2.111)$$

The corresponding magnitudes are

$$V_1 = [\dot{\vec{R}}^2]^{1/2} \qquad (2.112)$$

$$V_2 = [\dot{\vec{R}}^2 + \dot{\vec{L}}_1^2 + 2\dot{\vec{R}} \cdot \dot{\vec{L}}_1]^{1/2} \qquad (2.113)$$

$$V_3 = [V_2^2 + 2\dot{\vec{R}} \cdot \dot{\vec{L}}_2 + 2\dot{\vec{L}}_1 \cdot \dot{\vec{L}}_2]^{1/2} \qquad (2.114)$$

The $\dot{\vec{R}}$, $\dot{\vec{L}}_1$, and $\dot{\vec{L}}_2$ can be written as

$$\dot{\vec{R}} = \left(\dot{\vec{R}}\right)_{x_oy_oz_o} + \vec{\omega}_o \times \vec{R} \qquad (2.115)$$

$$\dot{\vec{L}}_1 = \left(\dot{\vec{L}}_1\right)_{x_1y_1z_1} + \vec{\omega}_1 \times \vec{L}_1 \qquad (2.116)$$

$$\dot{\vec{L}}_2 = \left(\dot{\vec{L}}_2\right)_{x_2y_2z_2} + \vec{\omega}_2 \times \vec{L}_2 \qquad (2.117)$$

Knowing that the system is in a circular orbit, and the cable connecting the two spacecraft is moving with a constant speed of v, we get

$$\left(\dot{\vec{R}}\right)_{x_oy_oz_o} = 0, \quad \left(\dot{\vec{L}}_1\right)_{x_1y_1z_1} = 0, \quad \left(\dot{\vec{L}}_2\right)_{x_2y_2z_2} = 0 \qquad (2.118)$$

Substituting the above relations in Eqs. (2.115)-(2.117), we obtain

$$\dot{\vec{R}} = \vec{\omega}_o \times \vec{R}, \quad \dot{\vec{L}}_1 = \vec{\omega}_1 \times \vec{L}_1, \quad \dot{\vec{L}}_2 = \vec{\omega}_2 \times \vec{L}_2 \qquad (2.119)$$

The terms $\dot{\vec{R}}^2$, $\dot{\vec{L}}_1^2$, and $\dot{\vec{L}}_2^2$ can be written as

$$\dot{\vec{R}}^2 = (\vec{\omega}_o \times \vec{R})^2, \quad \dot{\vec{L}}_1^2 = (\vec{\omega}_1 \times \vec{L}_1)^2, \quad \dot{\vec{L}}_2^2 = (\vec{\omega}_2 \times \vec{L}_2)^2 \qquad (2.120)$$

Writing $\vec{\omega}_o$, $\vec{R}$, $\vec{\omega}_1$, $\vec{L}_1$, and $\vec{\omega}_2$, and $\vec{L}_2$ in terms of the unit vectors of the respective coordinate frames, we have

$$\omega_o = \dot{\theta}\hat{k}_o, \quad \vec{R} = R\hat{i}_o, \quad \vec{\omega}_1 = (\dot{\theta} + \dot{\beta}_1)\hat{k}_1, \quad \vec{L}_1 = L_1\hat{i}_1 \vec{\omega}_2 = (\dot{\theta} + \dot{\beta}_2)\hat{k}_2,$$
$$\vec{L}_2 = L_2\hat{i}_2 \qquad (2.121)$$

The preceding angular velocity expressions can be further written as

$$\omega_o = \omega_o\hat{k}_o, \quad \vec{\omega}_1 = \omega_1\hat{k}_1, \quad \vec{\omega}_2 = \omega_2\hat{k}_2 \qquad (2.122)$$

where

$$\omega_o = \dot{\theta}, \quad \omega_1 = \dot{\theta} + \dot{\beta}_1, \quad \omega_2 = \dot{\theta} + \dot{\beta}_2 \qquad (2.123)$$

Inserting these expressions into Eqs. (2.120) and solving, we have

$$\dot{\vec{R}}^2 = \omega_o^2 R^2, \quad \dot{\vec{L}}_1^2 = \omega_1^2 L_1^2, \quad \dot{\vec{L}}_2^2 = \omega_2^2 L_2^2 \qquad (2.124)$$

Next we derive $\dot{\vec{R}} \cdot \dot{\vec{L}}_1$, $\dot{\vec{R}} \cdot \dot{\vec{L}}_2$ and $\dot{\vec{L}}_1 \cdot \dot{\vec{L}}_2$. Using Eqs. (2.119), we can write

$$\dot{\vec{R}} \cdot \dot{\vec{L}}_1 = (\vec{\omega}_o \times \vec{R}) \cdot (\vec{\omega}_1 \times \vec{L}_1) = \omega_o\omega_1 RL_1(\hat{j}_o \cdot \hat{j}_1) \qquad (2.125)$$

Similarly, we can derive

$$\dot{\vec{R}} \cdot \dot{\vec{L}}_2 = (\vec{\omega}_o \times \vec{R}) \cdot (\vec{\omega}_2 \times \vec{L}_2) = \omega_o\omega_2 RL_2(\hat{j}_o \cdot \hat{j}_2) \qquad (2.126)$$

$$\dot{\vec{L}}_1 \cdot \dot{\vec{L}}_2 = (\vec{\omega}_1 \times \vec{L}_1) \cdot (\vec{\omega}_2 \times \vec{L}_2) = \omega_1\omega_2 L_1 L_2(\hat{j}_1 \cdot \hat{j}_2) \qquad (2.127)$$

From the coordinate transformation between the coordinate frames $S - \hat{i}_o\hat{j}_o\hat{k}_o$ and $S - \hat{i}_1\hat{j}_1\hat{k}_1$, and between the coordinate frames $S - \hat{i}_o\hat{j}_o\hat{k}_o$ and $S - \hat{i}_2\hat{j}_2\hat{k}_2$, we have

$$\hat{j}_o \cdot \hat{j}_1 = cos\beta_1, \quad \hat{j}_o \cdot \hat{j}_2 = cos\beta_2, \quad \hat{j}_1 \cdot \hat{j}_2 = cos(\beta_2 - \beta_1) \qquad (2.128)$$

Thus, we can express $\dot{\vec{R}} \cdot \dot{\vec{L}}_1$, $\dot{\vec{R}} \cdot \dot{\vec{L}}_2$, and $\dot{\vec{L}}_1 \cdot \dot{\vec{L}}_2$ as

$$\dot{\vec{R}} \cdot \dot{\vec{L}}_1 = \omega_o\omega_1 RL_1 cos\beta_1 \qquad (2.129)$$

$$\dot{\vec{R}} \cdot \dot{\vec{L}}_2 = \omega_o\omega_2 RL_2 cos\beta_2 \qquad (2.130)$$

$$\dot{\vec{L}}_1 \cdot \dot{\vec{L}}_2 = \omega_1\omega_2 L_1 L_2 cos(\beta_2 - \beta_1) \qquad (2.131)$$

Substituting the expressions for $\dot{\vec{R}}^2$, $\dot{\vec{L}}_1^2$, and $\dot{\vec{L}}_2^2$ from Eqs. (2.124) and the expressions for $\dot{\vec{R}} \cdot \dot{\vec{L}}_1$, $\dot{\vec{R}} \cdot \dot{\vec{L}}_2$, and $\dot{\vec{L}}_1 \cdot \dot{\vec{L}}_2$ from Eqs. (2.131) into Eqs. (2.113)-(2.114), we finally obtain the magnitudes of the inertial velocity vectors for spacecraft m_1, m_2, and m_3 as

$$V_1 = \omega_o R \tag{2.132}$$

$$V_2 = [\omega_o^2 R^2 + \omega_1^2 L_1^2 + 2\omega_o\omega_1 R L_1 cos\beta_1]^{1/2} \tag{2.133}$$

$$V_3 = [V_2^2 + \omega_2^2 L_2^2 + 2\omega_o\omega_2 R L_2 cos\beta_2 + 2\omega_1\omega_2 L_1 L_2 cos(\beta_2 - \beta_1)]^{1/2} \tag{2.134}$$

The inertial acceleration vectors for the spacecraft m_1, m_2, and m_3 are written using Eqs. (2.110)-(2.111) for their velocity vectors, as

$$\vec{a}_1 = \dot{\vec{V}}_1 = \ddot{\vec{R}} \tag{2.135}$$

$$\vec{a}_2 = \dot{\vec{V}}_2 = \ddot{\vec{R}} + \ddot{\vec{L}}_1 \tag{2.136}$$

$$\vec{a}_3 = \dot{\vec{V}}_3 = \ddot{\vec{R}} + \ddot{\vec{L}}_1 + \ddot{\vec{L}}_2 \tag{2.137}$$

Here $\ddot{\vec{R}}$, $\ddot{\vec{L}}_1$, and $\ddot{\vec{L}}_2$ can be expressed as

$$\ddot{\vec{R}} = \ddot{\vec{R}}_{x_oy_oz_o} + 2(\vec{\omega}_o \times \dot{\vec{R}}_{x_oy_oz_o}) + \vec{\omega}_o \times (\vec{\omega} \times \vec{R}) + \dot{\vec{\omega}}_o \times \vec{R} \tag{2.138}$$

$$\ddot{\vec{L}}_1 = \left(\ddot{\vec{L}}_1\right)_{x_1y_1z_1} + 2\left[\vec{\omega}_1 \times \left(\dot{\vec{L}}_1\right)_{x_1y_1z_1}\right] + \vec{\omega}_1 \times (\vec{\omega}_1 \times \vec{L}_1) + \dot{\vec{\omega}}_1 \times \vec{L}_1 \tag{2.139}$$

$$\ddot{\vec{L}}_2 = \left(\ddot{\vec{L}}_2\right)_{x_2y_2z_2} + 2\left[\vec{\omega}_2 \times \left(\dot{\vec{L}}_2\right)_{x_2y_2z_2}\right] + \vec{\omega}_2 \times (\vec{\omega}_2 \times \vec{L}_2) + \dot{\vec{\omega}}_2 \times \vec{L}_2 \tag{2.140}$$

Knowing that the system is in a circular orbit and the cable connecting the two spacecraft is constant, *i.e.*,

$$\dot{\vec{R}}_{x_oy_oz_o} = \ddot{\vec{R}}_{x_oy_oz_o} = 0, \quad \dot{\vec{\omega}}_o = 0, \quad \left(\ddot{\vec{L}}_1\right)_{x_1y_1z_1} = 0, \quad \left(\ddot{\vec{L}}_2\right)_{x_2y_2z_2} = 0 \tag{2.141}$$

and writing all vectors in terms of the unit vectors along the respective coordinate frames, we obtain

$$\ddot{\vec{R}} = -\omega_o^2 R \hat{i}_o \tag{2.142}$$

$$\ddot{\vec{L}}_1 = -\omega_1^2 L_1 \hat{i}_1 + \ddot{\beta}_1 L_1 \hat{j}_1 \tag{2.143}$$

Similarly,

$$\ddot{\vec{L}}_2 = -\omega_2^2 L_2 \hat{i}_2 + \ddot{\beta}_2 L_2 \hat{j}_2 \tag{2.144}$$

Substituting the preceding relations in Eqs. (2.135)-(2.137), we have

$$\vec{a}_1 = -\omega_o^2 R \hat{i}_o \tag{2.145}$$

$$\vec{a}_2 = -\omega_o^2 R \hat{i}_o - \omega_1^2 L_1 \hat{i}_1 + \ddot{\beta}_1 L_1 \hat{j}_1 \tag{2.146}$$

$$\vec{a}_3 = \vec{a}_2 + \ddot{\vec{L}}_2 = \vec{a}_2 - \omega_2^2 L_2 \hat{i}_2 + \ddot{\beta}_2 L_2 \hat{j}_2 \tag{2.147}$$

The corresponding magnitude for spacecraft m_2 can be expressed as

$$a_2 = \left\{ (\omega_o^2 R)^2 + \omega_1^4 L_1^2 + \ddot{\beta}_1^2 L_1^2 + 2\omega_o^2 \omega_1^2 R L_1 (\hat{i}_o \cdot \hat{i}_1) - 2\omega_o^2 \ddot{\beta}_1 L_1 R (\hat{i}_o \cdot \hat{j}_1) \right\}^{1/2} \tag{2.148}$$

$$
\begin{aligned}
a_3 =& \left[a_2^2 + \omega_2^4 L_2^2 + \ddot{\beta}_2^2 L_2^2 + 2 L_2 \vec{a}_2 \cdot (-\omega_2^2 \hat{i}_2 + \ddot{\beta}_2 \hat{j}_2) \right]^{1/2} \\
=& \left\{ a_2^2 + \omega_2^4 L_2^2 + \ddot{\beta}_2^2 L_2^2 + 2\omega_o^2 \omega_2^2 R L_2 (\hat{i}_o \cdot \hat{i}_2) - 2\omega_o^2 \ddot{\beta}_2 R L_2 (\hat{i}_o \cdot \hat{j}_2) \right. \\
& \left. + 2\omega_1^2 \omega_2^2 L_1 L_2 (\hat{i}_1 \cdot \hat{i}_2) - 2\omega_1^2 \ddot{\beta}_2 L_1 L_2 (\hat{i}_1 \cdot \hat{j}_2) - 2\omega_2^2 L_1 L_2 (\hat{j}_1 \cdot \hat{i}_2) \right. \\
& \left. + 2\ddot{\beta}_1 \ddot{\beta}_2 L_1 L_2 (\hat{j}_1 \cdot \hat{j}_2) \right\}^{1/2}
\end{aligned}
\tag{2.149}
$$

From the coordinate transformation between the coordinate frames $S - \hat{i}_o \hat{j}_o \hat{k}_o$ and $S - \hat{i}_1 \hat{j}_1 \hat{k}_1$, we get

$$\hat{i}_o \cdot \hat{i}_1 = cos\beta_1, \quad \hat{i}_o \cdot \hat{j}_1 = cos(90 + \beta_1) = -sin\beta_1, \quad \hat{i}_o \cdot \hat{i}_2 = cos\beta_2,$$

$$\hat{i}_o \cdot \hat{j}_2 = cos(90 + \beta_2) = -sin\beta_2, \quad \hat{i}_1 \cdot \hat{i}_2 = cos(\beta_2 - \beta_1),$$

$$\hat{i}_1 \cdot \hat{j}_2 = cos(90 + \beta_2 - \beta_1) = -sin(\beta_2 - \beta_1),$$

$$\hat{j}_1 \cdot \hat{i}_2 = cos(90 - (\beta_2 - \beta_1)) = sin(\beta_2 - \beta_1), \quad \hat{j}_1 \cdot \hat{j}_2 = cos(\beta_2 - \beta_1) \tag{2.150}$$

Using the preceding relations into Eqs. (2.148), we obtain the accelerations as

$$a_2 = \left\{ (\omega_o^2 R)^2 + \omega_1^4 L_1^2 + \ddot{\beta}_1^2 L_1^2 + 2\omega_o^2 \omega_1^2 R L_1 cos\beta_1 + 2\omega_o^2 \ddot{\beta}_1 L_1 R sin\beta_1 \right\}^{1/2} \tag{2.151}$$

$$
\begin{aligned}
a_3 =& \left\{ a_2^2 + \omega_2^4 L_2^2 + \ddot{\beta}_2^2 L_2^2 + 2\omega_o^2 \omega_2^2 R L_2 cos\beta_2 - 2\omega_o^2 \ddot{\beta}_2 R L_2 sin\beta_2 \right. \\
& \left. + 2\omega_1^2 \omega_2^2 L_1 L_2 cos(\beta_2 - \beta_1) + 2\omega_1^2 \ddot{\beta}_2 L_1 L_2 sin(\beta_2 - \beta_1) \right. \\
& \left. + 2\omega_2^2 \ddot{\beta}_1 L_1 L_2 sin(\beta_2 - \beta_1) + 2\ddot{\beta}_1 \ddot{\beta}_2 L_1 L_2 cos(\beta_2 - \beta_1) \right\}^{1/2}
\end{aligned}
\tag{2.152}
$$

2.6 Let $\vec{r}_1$, $\vec{r}_2$, and $\vec{r}_3$ denote the position vectors of m_1, m_2, and m_3, respectively. We can write the center of mass relation as

$$m_1 \vec{r}_1 + m_2 \vec{r}_2 + m_3 \vec{r}_3 = 0 \tag{2.153}$$

Knowing

$$\vec{r}_2 = \vec{r}_1 + \vec{L}_1, \quad \vec{r}_3 = \vec{r}_2 + \vec{L}_2 \tag{2.154}$$

we have

$$\vec{r}_1 = -\gamma_1 L_1 \hat{i}_1 - \gamma_2 L_2 \hat{i}_2 \tag{2.155}$$

$$\vec{r}_2 = (1 - \gamma_1) L_1 \hat{i}_1 - \gamma_2 L_2 \hat{i}_2 \tag{2.156}$$

$$\vec{r}_3 = (1 - \gamma_1) L_1 \hat{i}_1 + (1 - \gamma_2) L_2 \hat{i}_2 \tag{2.157}$$

where

$$\gamma_1 = \frac{m_2 + m_3}{M}$$

$$\gamma_2 = \frac{m_3}{M}$$

$$M = m_1 + m_2 + m_3$$

2.13 The inertial position of body m_1 is

$$\vec{R}_1 = \vec{R} + \vec{r}_1 \tag{2.158}$$

The corresponding inertial acceleration is

$$\ddot{\vec{R}}_1 = \ddot{\vec{R}} + \ddot{\vec{r}}_1 \tag{2.159}$$

where $\ddot{\vec{R}}$ and $\ddot{\vec{x}}$ are derived next.

From the center of mass relation, we can write

$$m_1 \vec{r}_1 + m_2 \vec{r}_2 + \rho x \left(\vec{r}_1 - \frac{\vec{x}}{2} \right) + \rho(L - x) \left(\vec{r}_1 + \frac{\vec{L} - \vec{x}}{2} \right) = 0 \tag{2.160}$$

Knowing $\vec{r}_2 = \vec{r}_1 + \vec{L} - \vec{x}$, we solve for $\vec{r}_1$ and obtain

$$\vec{r}_1 = -\frac{1}{M} \left[\left(m_2 + \frac{m_L}{2} \right) \vec{L} - (m_2 + m_L) \vec{x} \right] \tag{2.161}$$

Here $m_L = \rho L$. Note $\rho x \vec{L}/2 = \rho L \vec{x}/2$ as $\vec{L} \parallel \vec{x}$.

Taking $\gamma_1 = -(m_2 + m_L/2)/M$ and $\gamma_2 = (m_2 + m_L)/M$, we can rewrite $\vec{r}_1$ as

$$\vec{r}_1 = \gamma_1 \vec{L} + \gamma_2 \vec{x} \tag{2.162}$$

Differentiating twice and applying the relation for inertial acceleration,

$$\ddot{\vec{r}}_{XYZ} = \ddot{\vec{r}}_{xyz} + 2(\vec{\omega} \times \dot{\vec{r}}_{xyz}) + \vec{\omega} \times (\vec{\omega} \times \vec{r}_{xyz}) + \dot{\vec{\omega}} \times \vec{r}_{xyz}$$

we obtain

$$\ddot{\vec{r}}_1 = \gamma_1 \ddot{\vec{L}} + \gamma_2 \ddot{\vec{x}}$$
$$= \gamma_1[-\omega^2 L\hat{i} + \dot{\beta}L\hat{j}] + \gamma_2[\ddot{x}\hat{i} + 2\omega\dot{x}\hat{j} - \omega^2 x\hat{i} + \dot{\beta}x\hat{j}]$$
$$= [-\gamma_1\omega^2 L + \gamma_2\ddot{x} - \gamma_2\omega^2 x]\hat{i} + [\gamma_1\dot{\beta}L + 2\gamma_2\omega\dot{x} + \gamma_2\dot{\beta}x]\hat{j} \quad (2.163)$$

where $\vec{\omega} = (\dot{\theta} + \dot{\beta})\hat{k}$ and $\dot{\vec{\omega}} = \ddot{\beta}$ as $\ddot{\theta} = 0$.

We can further express $\ddot{\vec{r}}_1$ as

$$\ddot{\vec{r}}_1 = A\hat{i} + B\hat{j} \qquad (2.164)$$

where $A = [-\gamma_1\omega^2 L + \gamma_2\ddot{x} - \gamma_2\omega^2 x]$ and $B = [\gamma_1\dot{\beta}L + 2\gamma_2\omega\dot{x} + \gamma_2\dot{\beta}x]$.

The term $\ddot{\vec{R}}$ can be expressed as

$$\ddot{\vec{R}} = -R\dot{\theta}^2\hat{i}_o = C\hat{i}_o \qquad (2.165)$$

where $C = -R\dot{\theta}^2$.

Substituting Eqs. (2.164) and (2.165) into Eq. (2.159), we obtain the inertial acceleration of m_1 as

$$\ddot{\vec{R}}_1 = C\hat{i}_o + A\hat{i} + B\hat{j} \qquad (2.166)$$

The corresponding magnitude is

$$a_1 = |\ddot{\vec{R}}_1| = \left[C^2 + A^2 + B^2 + 2CA(\hat{i}_o \cdot \hat{i}) + 2CB(\hat{i}_o \cdot \hat{j}) + 2AB(\hat{i} \cdot \hat{j})\right]^{1/2} \qquad (2.167)$$

Using coordinate transformation between $S - \hat{i}_o\hat{j}_o\hat{k}_o$ and $S - \hat{i}\hat{j}\hat{k}$:

$$(\hat{i}_o \cdot \hat{i}) = cos\beta \qquad (2.168)$$
$$(\hat{i}_o \cdot \hat{j}) = -sin\beta \qquad (2.169)$$

and knowing $(\hat{i} \cdot \hat{j}) = 0$, we have

$$a_1 = \left[C^2 + A^2 + B^2 + 2CAcos\beta - 2CBsin\beta\right]^{1/2} \qquad (2.170)$$

If $m_1 \ll m_2$, then mass ratios γ_1 and γ_2 reduce to

$$\gamma_1 = -1, \quad \gamma_2 = 1 \qquad (2.171)$$

Applying these values, A and B can be rewritten as

$$A = \omega^2(L - x) + \ddot{x}, \quad B = -\ddot{\beta}(L - x) + 2\omega\dot{x} \qquad (2.172)$$

Substituting these expressions in Eq. (2.170), we can obtain the inertial acceleration of m_1 when $m_1 \ll m_2$.

Considering the case of $m_1 \ll m_2$, and assuming $\ddot{\beta} = \ddot{x} = \dot{x} = 0$, we have

$$A = \omega^2(L - x), \quad B = 0 \tag{2.173}$$

and the corresponding inertial acceleration becomes

$$a_1 = \left[C^2 + A^2 + B^2 + 2CA\cos\beta\right]^{1/2} \tag{2.174}$$

From the preceding equation, the acceleration has maximum and minimum values at $\beta = 0$ and $\beta = \pi$, respectively. However, in the case of librating system with $\beta \leq \pi/2$, the minimum acceleration occurs at $\beta = \pi/2$.

2.14 The linear momentum of the system is

$$\vec{p} = (m_1 + m_2 + m_L)\dot{\vec{R}} = M\omega_o\vec{R} \tag{2.175}$$

where $M = m_1 + m_2 + m_L$ and $m_L = \rho L$.

The angular momentum of the system is given by

$$\begin{aligned}
\vec{H} = &(m_1 + m_2 + m_L)(\vec{R} \times \dot{\vec{R}}) + m_2(\vec{r}_2 \times \dot{\vec{r}}_2) \\
&+ m_L(\vec{r}_L \times \dot{\vec{r}}_L) + I\vec{\omega} + I_L\vec{\omega}_L
\end{aligned} \tag{2.176}$$

where $\vec{r}_2 = \vec{a} + \vec{L}$ and $\vec{r}_L = \vec{a} + \vec{L}/2$.

We can further write the angular momentum of the system as

$$\begin{aligned}
\vec{H} = &(m_1 + m_2 + m_L)R^2\vec{\omega}_o + (m_L + m_2)[a^2\vec{\omega} + a(L/2)\cos(\alpha - \beta)\vec{\omega}_L] \\
&+ (m_L/2 + m_2)\{aL\cos(\alpha - \beta)\vec{\omega} + aL\cos(\alpha - \beta)\vec{\omega}_L\} \\
&+ (m_L/4 + m_2)L^2\vec{\omega}_L + (I\omega + I_L\omega_L)\hat{k}
\end{aligned} \tag{2.177}$$

where

$$\omega_o = \dot{\theta}, \quad \omega = \dot{\theta} + \dot{\alpha}, \quad \omega_L = \dot{\theta} + \dot{\beta}$$
$$I_L = \frac{1}{12}m_LL^2$$

2.15 The potential energy U of the system is sum of the potential energy due to the spacecraft m_1, U_1 and the potential energy due to the spacecraft m_2, U_2, *i.e.*,

$$U = U_1 + U_2 \tag{2.178}$$

The potential energies due to the spacecraft m_1 and m_2 are

$$U_1 = -\frac{\mu m_1}{R}\left[1 - \frac{\vec{R}\cdot\vec{r}_1}{R^2} - \frac{1}{2}\frac{r_1^2}{R^2} + \frac{3}{2}\frac{(\vec{R}\cdot\vec{r}_1)^2}{R^4}\right] \qquad (2.179)$$

$$U_2 = -\frac{\mu m_2}{R}\left[1 - \frac{\vec{R}\cdot\vec{r}_2}{R^2} - \frac{1}{2}\frac{r_2^2}{R^2} + \frac{3}{2}\frac{(\vec{R}\cdot\vec{r}_2)^2}{R^4}\right] \qquad (2.180)$$

Thus, we can write the system potential energy U as per Eq.(2.178) using Eqs.(2.179-2.180) as

$$U = -\frac{\mu}{R}\left[(m_1 + m_2) - \frac{m_1(\vec{R}\cdot\vec{r}_1) + m_2(\vec{R}\cdot\vec{r}_2)}{R^2} - \frac{1}{2}\frac{m_1 r_1^2 + m_2 r_2^2}{R^2}\right.$$
$$\left. + \frac{3}{2}\frac{m_1(\vec{R}\cdot\vec{r}_1)^2 + m_2(\vec{R}\cdot\vec{r}_2)^2}{R^4}\right] \qquad (2.181)$$

As the center of mass lies at S, we get

$$m_1\vec{r}_1 + m_2\vec{r}_2 = 0 \qquad (2.182)$$

Let us consider the distance between m_1 and m_2 be L. Then

$$\vec{L} = \vec{r}_2 - \vec{r}_1 \qquad (2.183)$$

Using Eqs. (2.182-2.183), we can write $\vec{r}_1$ and $\vec{r}_2$ in terms of $\vec{L}$ as

$$\vec{r}_1 = -\frac{m_2}{m_1 + m_2}\vec{L} \qquad (2.184)$$

$$\vec{r}_2 = \frac{m_1}{m_1 + m_2}\vec{L} \qquad (2.185)$$

Now, we define $\vec{R}$ and $\vec{L}$ with respect to the orbital coordinate frame $S-x_o y_o z_o$ and the dumbbell fixed coordinate frame $S-xyz$, respectively as

$$\vec{R} = R\hat{i}_o; \quad \vec{L} = L\hat{i} \qquad (2.186)$$

Applying Eq. (2.182) and Eqs. (2.184-2.185) into Eq.(2.181), we have

$$U = -\frac{\mu M}{R} + \frac{\mu}{2R^3}M_e[1 - 3(\vec{i}_o \cdot \vec{i})^2]L^2 \qquad (2.187)$$

where M and M_e denote system mass and equivalent system mass, respectively. They are

$$M = m_1 + m_2 \qquad (2.188)$$

$$M_e = \frac{m_1 m_2}{m_1 + m_2} \qquad (2.189)$$

The transformation from the frame $S - x_o y_o z_o$ to the frame $S - xyz$ is obtained by a rotation of β about x_o-axis. We have

$$\left\{ \begin{array}{c} \hat{i} \\ \hat{j} \\ \hat{k} \end{array} \right\} = R_z y(\beta, \eta) \left\{ \begin{array}{c} \hat{i}_o \\ \hat{j}_o \\ \hat{k}_o \end{array} \right\} \qquad (2.190)$$

where $R_{zy}(\beta, \eta)$ is

$$R_{zy}(\beta, \eta) = R_z(\eta) R_z(\beta) = \begin{bmatrix} cos\eta & 0 & -sin\eta \\ 0 & 1 & 0 \\ sin\eta & 0 & cos\eta \end{bmatrix} \begin{bmatrix} cos\beta & sin\beta & 0 \\ -sin\beta & cos\beta & 0 \\ 0 & 0 & 1 \end{bmatrix}$$

$$= \begin{bmatrix} cos\beta cos\eta & sin\beta cos\eta & -sin\eta \\ -sin\beta & cos\beta & 0 \\ cos\beta sin\eta & sin\beta sin\eta & cos\eta \end{bmatrix} \qquad (2.191)$$

Using Eq.(2.190), we can write $\hat{i}$ as

$$\hat{i} = cos\beta cos\eta \hat{i}_o + sin\beta cos\eta \hat{j}_o - sin\eta \hat{k}_o \qquad (2.192)$$

Thus, we obtain $(\hat{i}_o \cdot \hat{i})$ as

$$\hat{i}_o \cdot \hat{i} = cos\beta cos\eta \qquad (2.193)$$

and substituting it in Eq. (2.187), we get the system potential energy

$$U = -\frac{\mu M}{R} + \frac{\mu}{2R^3} M_e (1 - 3cos^2\beta cos^2\eta) L^2 \qquad (2.194)$$

To obtain the maximum and minimum values of the system potential energy, we differentiate the preceding relation and equate it to zero as follows:

$$\frac{dU}{d\beta} = 0$$

$$\Rightarrow -\frac{6\mu}{2R^3} (M_e cos\beta sin\beta cos^2\eta) L^2 = 0 \qquad (2.195)$$

or

$$cos\beta sin\beta cos^2\eta = 0 \tag{2.196}$$

Knowing $\eta = 0$, we have

$$sin2\beta = 0 \tag{2.197}$$

so,

$$2\beta = 0, \pi, 2\pi, 3\pi, \cdots$$
$$\Rightarrow \beta = 0, \frac{\pi}{2}, \pi, \frac{3\pi}{2}, \cdots \tag{2.198}$$

Substituting $\beta = 0$ in the potential energy expression (2.194), we obtain

$$U_{\beta=0} = -\frac{\mu M}{R} + \frac{\mu}{2R^3} M_e (1 - 3cos^2\eta)L^2 \tag{2.199}$$

The above result remains same for $\beta = \pi, 2\pi, 3\pi, \cdots$.

In the case of $\beta = \pi/2$, the potential energy is

$$U_{\beta=\pi/2} = -\frac{\mu M}{R} + \frac{\mu}{2R^3} M_e L^2 \tag{2.200}$$

The preceding result remains same for $\beta = (3/2)\pi, (5/2)\pi, \cdots$.

Comparing Eqs. (2.199) and (2.200), we find

$$U_{\beta=\pi/2} > U_{\beta=0} \tag{2.201}$$

Thus, the system potential energy is minimum at $\beta = 0, \pi, 2\pi, 3\pi, \cdots$ and maximum at $\beta = (1/2)\pi, (3/2)\pi, (5/2)\pi, \cdots$.

2.16 Using *Summary Sheet (System: Three Point Masses)*, we can write the kinetic and potential energies of the system ($N = 3$) as

$$T = \frac{1}{2} \sum_{i=1}^{3} m_i \dot{\vec{R}}^2 + \frac{1}{2} \sum_{i=1}^{3} m_i \dot{\vec{r}}_i^2 \tag{2.202}$$

$$U = -\frac{\mu}{R} \sum_{i=1}^{3} m_i + \frac{\mu}{2R^3} \sum_{i=1}^{3} m_i r_i^2$$
$$- \frac{3\mu}{2R^5} \sum_{i=1}^{3} m_i (\vec{R} \cdot \vec{r}_i)^2 \tag{2.203}$$

The position vectors of the masses and the corresponding velocity vectors are

$$\vec{r}_1 = -\gamma_1 \vec{L}_1 - \gamma_2 \vec{L}_2, \quad \vec{r}_2 = (1 - \gamma_1)\vec{L}_1 - \gamma_2 \vec{L}_2$$
$$\vec{r}_3 = (1 - \gamma_1)\vec{L}_1 + (1 - \gamma_2)\vec{L}_2 \tag{2.204}$$

$$\dot{\vec{r}}_1 = -\gamma_1 \dot{\vec{L}}_1 - \gamma_2 \dot{\vec{L}}_2, \quad \dot{\vec{r}}_2 = (1 - \gamma_1)\dot{\vec{L}}_1 - \gamma_2 \dot{\vec{L}}_2$$
$$\dot{\vec{r}}_3 = (1 - \gamma_1)\dot{\vec{L}}_1 + (1 - \gamma_2)\dot{\vec{L}}_2 \tag{2.205}$$

where $\gamma_1 = (m_2 + m_3)/M$, $\gamma_2 = m_3/M$ and $M = m_1 + m_2 + m_3$. Squaring the preceding equations we have

$$\dot{\vec{r}}_1^2 = \gamma_1^2 \dot{\vec{L}}_1^2 + \gamma_2^2 \dot{\vec{L}}_2^2 + 2\gamma_1\gamma_2 \left(\dot{\vec{L}}_1 \cdot \dot{\vec{L}}_2 \right) \tag{2.206}$$

$$\dot{\vec{r}}_2^2 = (1 - \gamma_1)^2 \dot{\vec{L}}_1^2 + \gamma_2^2 \dot{\vec{L}}_2^2 - 2(1 - \gamma_1)\gamma_2 \left(\dot{\vec{L}}_1 \cdot \dot{\vec{L}}_2 \right) \tag{2.207}$$

$$\dot{\vec{r}}_3^2 = (1 - \gamma_1)^2 \dot{\vec{L}}_1^2 + (1 - \gamma_2)^2 \dot{\vec{L}}_2^2 + 2(1 - \gamma_1)(1 - \gamma_2) \left(\dot{\vec{L}}_1 \cdot \dot{\vec{L}}_2 \right) \tag{2.208}$$

Knowing

$$\dot{\vec{L}}_1 = \vec{\omega}_1 \times \vec{L}_1 = \omega_1 \hat{k}_1 \times L_1 \hat{i}_1 = \omega_1 L_1 \hat{j}_1 \tag{2.209}$$

$$\dot{\vec{L}}_2 = \vec{\omega}_2 \times \vec{L}_2 = \omega_2 \hat{k}_2 \times L_2 \hat{i}_2 = \omega_2 L_2 \hat{j}_2 \tag{2.210}$$

we have

$$\dot{\vec{r}}_1^2 = \gamma_1^2 \omega_1^2 L_1^2 + \gamma_2^2 \omega_2^2 L_2^2 + 2\gamma_1\gamma_2\omega_1\omega_2 L_1 L_2 cos(\beta_1 - \beta_2) \tag{2.211}$$

$$\dot{\vec{r}}_2^2 = (1 - \gamma_1)^2 \omega_1^2 L_1^2 + \gamma_2^2 \omega_2^2 L_2^2 - 2(1 - \gamma_1)\gamma_2\omega_1\omega_2 L_1 L_2 cos(\beta_1 - \beta_2) \tag{2.212}$$

$$\dot{\vec{r}}_3^2 = (1 - \gamma_1)^2 \omega_1^2 L_1^2 + (1 - \gamma_2)^2 \omega_2^2 L_2^2 + 2(1 - \gamma_1)(1 - \gamma_2)$$
$$\times \omega_1\omega_2 L_1 L_2 cos(\beta_1 - \beta_2) \tag{2.213}$$

Taking $\vec{R} = R\hat{i}_o$, we write $(\hat{i}_o \cdot \vec{r}_j)$, $j = 1, 2, 3$ as

$$(\hat{i}_o \cdot \vec{r}_1) = [-\gamma_1 L_1 \hat{i}_1 - \gamma_2 L_2 \hat{i}_2] = -\gamma_1 L_1 cos\beta_1 - \gamma_2 L_2 cos\beta_2 \tag{2.214}$$

$$(\hat{i}_o \cdot \vec{r}_2) = [(1 - \gamma_1)L_1 \hat{i}_1 - \gamma_2 L_2 \hat{i}_2] = (1 - \gamma_1)L_1 cos\beta_1 - \gamma_2 L_2 cos\beta_2 \tag{2.215}$$

$$(\hat{i}_o \cdot \vec{r}_3) = [(1 - \gamma_1)L_1 \hat{i}_1 + (1 - \gamma_2)L_2 \hat{i}_2] = (1 - \gamma_1)L_1 cos\beta_1 + (1 - \gamma_2)L_2 cos\beta_2 \tag{2.216}$$

Substituting the preceding relations into Eqs. (2.202)-(2.203) yield the kinetic and potential energies of the system as

$$T = \frac{1}{2}M\dot{R}^2 + \frac{1}{2}[m_1\gamma_1^2 + m_2(1-\gamma_1)^2 + m_3(1-\gamma_1)^2]\omega_1^2 L_1^2$$

$$+ \frac{1}{2}[m_1\gamma_2^2 + m_2\gamma_2^2 + m_3(1-\gamma_2)^2]\omega_2^2 L_2^2$$

$$+ [m_1\gamma_1\gamma_2 - m_2(1-\gamma_1)\gamma_2 + m_3(1-\gamma_1)(1-\gamma_2)]\omega_1\omega_2 L_1 L_2 cos(\beta_1 - \beta_2)$$

$$(2.217)$$

$$U = -\frac{\mu M}{R} + \frac{\mu}{2R^3}\left\{ [m_1\gamma_1^2 + m_2(1-\gamma_1)^2 + m_3(1-\gamma_1)^2]L_1^2 \right.$$

$$+ [m_1\gamma_2^2 + m_2\gamma_2^2 + m_3(1-\gamma_2)^2]L_2^2$$

$$\left. + 2[m_1\gamma_1\gamma_2 - m_2(1-\gamma_1)\gamma_2 + m_3(1-\gamma_1)(1-\gamma_2)]L_1 L_2 cos(\beta_1 - \beta_2) \right\}$$

$$- \frac{3\mu}{2R^3}\left\{ [m_1\gamma_1^2 + m_2(1-\gamma_1)^2 + m_3(1-\gamma_1)^2]L_1^2 cos^2\beta_1 \right.$$

$$+ [m_1\gamma_2^2 + m_2\gamma_2^2 + m_3(1-\gamma_2)^2]L_2^2 cos^2\beta_2$$

$$\left. + 2[m_1\gamma_1\gamma_2 - m_2(1-\gamma_1)\gamma_2^2 + m_3(1-\gamma_1)(1-\gamma_2)]L_1 L_2 cos\beta_1 cos\beta_2 \right\}$$

$$(2.218)$$

Taking

$$M_{t1} = m_1\gamma_1^2 + m_2(1-\gamma_1)^2 + m_3(1-\gamma_1)^2 \qquad (2.219)$$

$$M_{t2} = m_1\gamma_2^2 + m_2\gamma_2^2 + m_3(1-\gamma_2)^2 \qquad (2.220)$$

$$M_{t3} = m_1\gamma_1\gamma_2 - m_2(1-\gamma_1)\gamma_2 + m_3(1-\gamma_1)(1-\gamma_2) \qquad (2.221)$$

the kinetic and potential energies of the system are

$$T = \frac{1}{2}M\dot{R}^2 + \frac{1}{2}M_{t1}\omega_1^2 L_1^2 + \frac{1}{2}M_{t2}\omega_2^2 L_2^2 + M_{t3}\omega_1\omega_2 L_1 L_2 cos(\beta_1 - \beta_2)$$

$$(2.222)$$

$$U = -\frac{\mu M}{R} + \frac{\mu}{2R^3}\left\{ M_{t1}L_1^2 + M_{t2}L_2^2 + 2M_{t3}L_1 L_2 cos(\beta_1 - \beta_2) \right\}$$

$$- \frac{3\mu}{2R^3}\left\{ M_{t1}L_1^2 cos^2\beta_1 + M_{t2}L_2^2 cos^2\beta_2 + 2M_{t3}L_1 L_2 cos\beta_1 cos\beta_2 \right\}$$

$$(2.223)$$

2.17 The kinetic energy of the system (taking $\vec{r}_1 = 0$) is

$$\vec{T} = \frac{1}{2}(m_1 + m_2 + m_3)\dot{\vec{R}}^2 + \frac{1}{2}m_2\dot{\vec{r}}_2^2 + \frac{1}{2}m_3\dot{\vec{r}}_3^2 + \frac{1}{2}I_z\omega^2$$

where $\dot{\vec{r}}_2 = \dot{\vec{d}}$ and $\dot{\vec{r}}_3 = \dot{\vec{d}} + \dot{\vec{L}} + \dot{\vec{x}}$. The nomenclature $\vec{d}$, $\vec{L}$, and $\vec{x}$ are expressed with respect to coordinate frames as

$$\vec{d} = a\hat{i} + b\hat{j}, \quad \vec{x} = x\vec{j}_c = x[sin\beta\hat{i} - cos\beta\hat{j}], \quad \vec{L} = L\hat{j} \qquad (2.224)$$

(Note $\vec{j}_c$ is unit vector along cable) and their derivatives with respect to time are obtained as

$$\dot{\vec{d}} = \dot{a}\hat{i} + \vec{\omega} \times \hat{d} = (\dot{a} - \omega b)\hat{i} + \omega a\hat{j} \qquad (2.225)$$

$$\dot{\vec{L}} = \vec{\omega} \times \vec{L} = \omega\hat{k} \times L\hat{j} = -\omega L\hat{i} \qquad (2.226)$$

$$\dot{\vec{x}} = \omega_c x[cos\beta\hat{i} + sin\beta\hat{j}] \qquad (2.227)$$

Here $\omega_c = \dot{\theta} + \dot{\alpha} + \dot{\beta}$. We obtain the system kinetic energy as

$$\begin{aligned}
T = &\frac{1}{2}M\omega_o^2R^2 + \frac{1}{2}I_z\omega^2 + \frac{1}{2}(m_2 + m_3)\left\{(\dot{a} - \omega b)^2 + \omega^2 a^2\right\} \\
&+ \frac{1}{2}m_3\left\{\omega^2L^2 + x^2\omega_c^2 - 2\omega L(\dot{a} - \omega b)\right. \\
&+ 2\omega_c x[(\dot{a} - \omega b)cos\beta + \omega a sin\beta] - \left.2\omega\omega_c Lxcos\beta\right\} \qquad (2.228)
\end{aligned}$$

where $M = m_1 + m_2 + m_3$ and $\omega = \dot{\theta} + \dot{\alpha}$.

The system potential energy is

$$\begin{aligned}
U = -\frac{\mu}{R}&\left[(m_1 + m_2 + m_3) - \frac{1}{2}\frac{m_1r_1^2 + m_2r_2^2 + m_3r_3^2}{R^2}\right. \\
&+ \frac{3}{2}\frac{m_1(\vec{R} \cdot \vec{r}_1)^2 + m_2(\vec{R} \cdot \vec{r}_2)^2 + m_3(\vec{R} \cdot \vec{r}_3)^2}{R^4}\right] \\
&+ \frac{\mu}{4R^3}\left\{(I_x + I_y + I_z) - 3[I_z + (I_y - I_x)cos2\alpha]\right\}
\end{aligned}$$

Knowing

$$\vec{r}_1 = 0 \qquad (2.229)$$

$$\vec{r}_2 = \vec{d} = a\hat{i} + b\hat{j} \qquad (2.230)$$

$$\vec{r}_3 = \vec{d} + \vec{L} + \vec{x} = (a + xsin\beta)\hat{i} + (b + L - xcos\beta)\hat{j} \qquad (2.231)$$

we have the system potential energy as

$$U = -\frac{\mu(m_1 + m_2 + m_3)}{R} + \frac{\mu}{2R^3}\left\{ m_2(a^2 + b^2) + m_3[a^2 + (b+L)^2 + x^2 \right.$$

$$\left. + 2axsin\beta - 2(b+L)xcos\beta] \right\} - \frac{3\mu}{2R^3}\left\{ m_2[acos\alpha - bsin\alpha]^2 \right.$$

$$\left. + m_3[acos\alpha - (b+L)sin\alpha + xsin(\alpha + \beta)]^2 \right\}$$

$$+ \frac{\mu}{4R^3}\left\{ (I_x + I_y + I_z) - 3[I_z + (I_y - I_x)cos2\alpha] \right\} \qquad (2.232)$$

2.18 For a 1-3-2 Euler angle rotation sequence, we obtain the rotation matrix, $R_{132}(\alpha, \phi, \gamma)$ as

$$R_{132}(\alpha, \phi, \gamma) = R_2(\gamma)R_3(\phi)R_1(\alpha)$$

$$= \begin{bmatrix} cos\phi cos\gamma & cos\alpha sin\phi cos\gamma + sin\alpha sin\gamma & sin\alpha sin\phi cos\gamma - cos\alpha sin\gamma \\ -sin\phi & cos\alpha cos\phi & sin\alpha cos\phi \\ cos\phi sin\gamma & cos\alpha sin\phi sin\gamma - sin\alpha cos\gamma & sin\alpha sin\phi sin\gamma + cos\alpha cos\gamma \end{bmatrix}$$

$$(2.233)$$

Thus, the transformation from the frame $S - i_o j_o k_o$ to the body fixed frame $S - ijk$ using 1-3-2 Euler angle rotation sequence is

$$\left\{ \begin{array}{c} \hat{i} \\ \hat{j} \\ \hat{k} \end{array} \right\} = R_{132}(\alpha, \phi, \gamma) \left\{ \begin{array}{c} \hat{i_o} \\ \hat{j_o} \\ \hat{k_o} \end{array} \right\} \qquad (2.234)$$

We can obtain the rotation matrix R_{231} by taking the inverse of the transformation R_{132}^{-1}. However, from the properties of a rotation matrix explained earlier, R_{132}^{-1} is the transpose of R_{132} (*i.e.*, R_{132}^T). Thus, we find R_{231} as

$$R_{231}(-\gamma, -\phi, -\alpha) = \begin{bmatrix} cos\phi cos\gamma & -sin\phi & cos\phi sin\gamma \\ cos\alpha sin\phi cos\gamma + sin\alpha sin\gamma & cos\alpha cos\phi & cos\alpha sin\phi sin\gamma - sin\alpha cos\gamma \\ sin\alpha sin\phi cos\gamma - cos\alpha sin\gamma & sin\alpha cos\phi & sin\alpha sin\phi sin\gamma + cos\alpha cos\gamma \end{bmatrix}$$

$$(2.235)$$

and thus, the transformation from the frame $S - ijk$ to the body fixed frame $S - i_o j_o k_o$ is

$$\left\{ \begin{array}{c} \hat{i}_o \\ \hat{j}_o \\ \hat{k}_o \end{array} \right\} = R_{231} \left\{ \begin{array}{c} \hat{i} \\ \hat{j} \\ \hat{k} \end{array} \right\} \tag{2.236}$$

Apart from these rotation matrices, we require angular velocity $\vec{\omega}$ of the spacecraft as well. The angular velocity $\vec{\omega}$ of the spacecraft can be expressed as

$$\vec{\omega} = \omega_x \hat{i} + \omega_y \hat{j} + \omega_z \hat{k} = \dot{\alpha}\hat{i}_o + \dot{\phi}\hat{k}_1 + \dot{\gamma}\hat{j} \tag{2.237}$$

Applying a transformation matrix for i_o and k_1, we get

$$\omega_x = \dot{\alpha}cos\phi \cos\gamma - \dot{\phi}sin\gamma$$
$$\omega_y = - \dot{\alpha}sin\phi + \dot{\gamma} \tag{2.238}$$
$$\omega_z = \dot{\alpha}cos\phi \sin\gamma + \dot{\phi}cos\gamma$$

$$\tag{2.239}$$

or, we can write

$$\left\{ \begin{array}{c} \omega_x \\ \omega_y \\ \omega_z \end{array} \right\} = \begin{bmatrix} cos\phi \cos\gamma & -sin\gamma & 0 \\ -sin\phi & 0 & 1 \\ cos\phi \sin\gamma & cos\gamma & 0 \end{bmatrix} \left\{ \begin{array}{c} \dot{\alpha} \\ \dot{\phi} \\ \dot{\gamma} \end{array} \right\} \tag{2.240}$$

It is to be noted that using a 1-3-2 Euler angle sequence and considering α, ϕ, γ to be the successive rotation angles, the singularity occurs at $\phi = \pm\pi/2$.

2.19 (a) The rotation matrix for γ about the y-axis is

$$R_2(\gamma) = \begin{bmatrix} cos\gamma & 0 & -sin\gamma \\ 0 & 1 & 0 \\ sin\gamma & 0 & cos\gamma \end{bmatrix}$$

The rotation matrix for ϕ about the z-axis is

$$R_3(\phi) = \begin{bmatrix} cos\phi & sin\phi & 0 \\ -sin\phi & cos\phi & 0 \\ 0 & 0 & 1 \end{bmatrix}$$

The rotation matrix for α about the x-axis is

$$R_1(\alpha) = \begin{bmatrix} 1 & 0 & 0 \\ 0 & cos\alpha & sin\alpha \\ 0 & -sin\alpha & cos\alpha \end{bmatrix}$$

(b) The rotation matrix $R_{231}(\gamma, \phi, \alpha)$ is derived as

$$R_{231} = R_1(\alpha)R_3(\phi)R_2(\gamma)$$

$$= \begin{bmatrix} cos\phi cos\gamma & sin\phi & -cos\phi sin\gamma \\ -cos\alpha sin\phi cos\gamma + sin\alpha sin\gamma & cos\alpha cos\phi & cos\alpha sin\phi sin\gamma + sin\alpha cos\gamma \\ sin\alpha sin\phi cos\gamma + cos\alpha sin\gamma & -sin\alpha cos\phi & sin\alpha sin\phi sin\gamma + cos\alpha cos\gamma \end{bmatrix}$$

(c) The rotation matrix $R_{132}(\alpha, \phi, \gamma)$ is obtained

$$R_{132} = R_2(\gamma)R_3(\phi)R_1(\alpha) = R_{231}^T(-\gamma, -\phi, -\alpha)$$

$$= \begin{bmatrix} cos\phi cos\gamma & cos\alpha sin\phi cos\gamma + sin\alpha sin\gamma & sin\alpha sin\phi cos\gamma - cos\alpha sin\gamma \\ -sin\phi & cos\alpha cos\phi & sin\alpha cos\phi \\ cos\phi sin\gamma & cos\alpha sin\phi sin\gamma - sin\alpha cos\gamma & sin\alpha sin\phi sin\gamma + cos\alpha cos\gamma \end{bmatrix}$$

If α, ϕ and γ are assumed to be very small *i.e.*, $cos\alpha = cos\phi = 1$, $sin\alpha = \alpha$, $sin\phi = \phi$, and $\alpha\phi = 0$, then R_{132} and R_{31} are

$$R_{231} = \begin{bmatrix} 1 & \phi & -\gamma \\ -\phi & 1 & \alpha \\ \gamma & -\alpha & 1 \end{bmatrix} R_{132}$$

Thus, the order of rotation does not matter if α, ϕ and γ are very small.

2.20 **(a)** Either 3 Euler angles plus the defined sequence or 4 Euler parameters (q_1, q_2, q_3, q_4).

(b) Yes. The Euler's equations of motion of a torque-free body are

$$I_1\dot{\omega}_1 - (I_2 - I_3)\omega_2\omega_3 = 0$$
$$I_2\dot{\omega}_2 - (I_3 - I_1)\omega_3\omega_1 = 0$$
$$I_3\dot{\omega}_3 - (I_1 - I_2)\omega_1\omega_2 = 0$$

For the conditions of $\omega_1 = \omega_0$, $\omega_2 = \omega_3 = 0$, and $\dot{\omega}_1 = 0$, we have the Euler's equations of motion

$$I_1 \times 0 = 0$$
$$I_2\dot{\omega}_2 = 0$$
$$I_3\dot{\omega}_3 = 0$$

Thus, ω_2=constant=0, and ω_3=constant=0.

(c) Using the Euler rotations the orientation of an aircraft can be specified completely by a sequence of three consecutive rotations about different aircraft body axes. The first and the last rotations about the same body axes are possible and thus, we have 12 such combinations *i.e.*, 1-2-3, 1-3-2, 2-1-3, 2-3-1, 3-1-2, 3-2-1, 2-1-2, 3-1-3, 1-2-1, 3-2-3, 1-3-1, and 2-3-2. Here, 1, 2, and 3 corresponds to x, y, and z axes, respectively. The rotation matrix about x, y, and z, are obtained as

$$R_1 = \begin{bmatrix} 1 & 0 & 0 \\ 0 & cos\theta & sin\theta \\ 0 & -sin\theta & cos\theta \end{bmatrix} \tag{2.241}$$

$$R_2 = \begin{bmatrix} cos\theta & 0 & -sin\theta \\ 0 & 1 & 0 \\ sin\theta & 0 & cos\theta \end{bmatrix} \tag{2.242}$$

$$R_3 = \begin{bmatrix} cos\theta & sin\theta & 0 \\ -sin\theta & cos\theta & 0 \\ 0 & 0 & 1 \end{bmatrix} \tag{2.243}$$

Considering 3-2-1 ($\psi - \theta - \phi$) Euler angle rotation sequence with transformation ($S - i_f j_f k_f \to S - i_1 j_1 k_1 \to S - i_2 j_2 k_2 \to S - ijk$), the aircraft angular velocity $\vec{\omega}$ can be written in the body-fixed frame

$$\vec{\omega} = p\hat{i} + q\hat{j} + r\hat{k} \tag{2.244}$$

and in the intermediate reference frame

$$\vec{\omega} = \dot{\psi}\hat{k}_f + \dot{\theta}\hat{j}_1 + \dot{\phi}\hat{i}_2 = \dot{\psi}\hat{k}_1 + \dot{\theta}\hat{j}_2 + \dot{\phi}\hat{i} \tag{2.245}$$

We have the following transformations:

$$\hat{k}_1 = -(sin\theta)\hat{i}_2 + (cos\theta)\hat{k}_2$$

$$\begin{Bmatrix} \hat{i}_2 \\ \hat{j}_2 \\ \hat{k}_2 \end{Bmatrix} = R_1^{-1} \begin{Bmatrix} \hat{i} \\ \hat{j} \\ \hat{k} \end{Bmatrix} = \begin{bmatrix} 1 & 0 & 0 \\ 0 & cos\phi & -sin\phi \\ 0 & sin\theta & cos\theta \end{bmatrix} \begin{Bmatrix} \hat{i} \\ \hat{j} \\ \hat{k} \end{Bmatrix}$$

where R_1^{-1} is the transpose of R_1.

Using these transformations, the angular velocity $\vec{\omega}$ can be written

$$\vec{\omega} = (-\dot{\psi}sin\theta + \dot{\phi})\hat{i} + (\dot{\psi}cos\theta sin\phi + \dot{\theta}cos\phi)\hat{j} + (\dot{\psi}cos\theta cos\phi)\hat{k}$$
$$\tag{2.246}$$

or

$$p = -\dot{\psi}sin\theta + \dot{\phi}$$
$$q = -\dot{\psi}cos\theta sin\phi + \dot{\theta}cos\phi \tag{2.247}$$
$$r = \dot{\psi}cos\theta \cos\phi - \dot{\theta}sin\phi$$

$$\tag{2.248}$$

or

$$\begin{Bmatrix} p \\ q \\ r \end{Bmatrix} = \begin{bmatrix} 1 & 0 & -sin\theta \\ 0 & cos\phi & cos\theta sin\phi \\ 0 & -sin\phi & cos\theta cos\phi \end{bmatrix} \begin{Bmatrix} \dot{\phi} \\ \dot{\theta} \\ \dot{\psi} \end{Bmatrix} \tag{2.249}$$

We solve the above equations for ϕ, θ, and ψ in terms of the Euler angles and rotational components. The determinant of the matrix on the right hand side is

$$
\begin{vmatrix}
1 & 0 & -sin\theta \\
0 & cos\phi & cos\theta sin\phi \\
0 & -sin\phi & cos\theta cos\phi
\end{vmatrix} = cos\theta cos^2\phi + cos\theta sin^2\phi = cos\theta
$$

$$(2.250)$$

The inverse of the above matrix does not exist if $\theta = \pm 90$ deg or we can say singularity occurs at $\theta = \pm 90$ deg. Using Euler angle sequences (1-3-2), (3-1-3), (3-1-2) the singularities occur at $\theta = \pm\pi/2$, $\theta=0$ or π, and $\theta = \pm\pi/2$, respectively. In fact, the determinant of the matrix (Eq. 2.250) may always involve sine or cosine terms of the the angle of the second rotation and therefore as sine or cosine angle becomes zero when the angle is 0 or π (sine angle) or $\pm\pi/2$ (cosine angle), the inverse of the determinant will have singularity. Therefore, no matter what sequence is taken for the Euler angle rotations, the angle of the second rotation displays a similar singularity at either zero or ± 90 deg.

To avoid the singularity problem, direction cosines or Euler parameters or quaternions are used to define the orientation of the spacecraft.

(d) Yes, it is possible to have the first and the last rotations about the same body axes in the Euler angle rotations?

2.21 The kinetic and potential energies of a rigid satellite are given by

$$
T = \frac{1}{2}m(\dot{R}^2 + \dot{\theta}^2 R^2) + \frac{1}{2}[I_x\omega_x^2 + I_y\omega_y^2 + I_z\omega_z^2] \qquad (2.251)
$$

$$
U = -\frac{\mu m}{R} - \frac{\mu}{4R^3}\Bigg\{ (I_{yy} + I_{zz} - I_{xx})[3(cos\alpha sin\phi cos\gamma + sin\alpha sin\gamma)^2 - 1]
$$
$$
+ (I_{zz} + I_{xx} - I_{yy})[3(cos\alpha cos\phi)^2 - 1]
$$
$$
+ (I_{xx} + I_{yy} - I_{zz})[3(cos\alpha sin\phi sin\gamma - sin\alpha cos\gamma)^2 - 1]\Bigg\}
$$

$$(2.252)$$

2.23

$$T = \frac{1}{2}M\dot{\vec{R}} + \frac{1}{2}(\omega^2 cos^2\eta + \dot{\eta}^2)M_e L^2 \qquad (2.253)$$

$$U = -\frac{\mu M}{R} + \frac{\mu}{2R^3}M_e(1 - 3cos^2\beta cos^2\eta)L^2 \qquad (2.254)$$

$$M_e = \frac{m_1(m_2 + m_L/2)^2}{M^2} + \frac{m_2(m_1 + m_L)^2}{M^2} + \frac{m_L(m_1 + m_L/2)^2}{M^2}$$
$$-\frac{m_L(m_1 + m_L)}{M} + \frac{m_L}{3} \qquad (2.255)$$

where $m_L = \rho L_0$, $M = m_1 + m_2 + m_L$, and $\omega = \dot{\theta} + \dot{\beta}$.

2.24

$$T = \frac{1}{2}M\dot{\vec{R}} + \frac{1}{2}M_e[\dot{u}^2 + (L_0 + u)^2\omega^2] \qquad (2.256)$$

$$U_e = \frac{1}{2}\frac{EA}{L_0}u^2 \qquad (2.257)$$

$$M_e = \frac{m_1(m_2 + m_L/2)^2}{M^2} + \frac{m_2(m_1 + m_L)^2}{M^2} + \frac{m_L(m_1 + m_L/2)^2}{M^2}$$
$$-\frac{m_L(m_1 + m_L)}{M} + \frac{m_L}{3} \qquad (2.258)$$

where $m_L = \rho L_0$, $M = m_1 + m_2 + m_L$, and $\omega = \dot{\theta} + \dot{\beta}$.

Chapter 3

Forces and Torques

Problem Set 3

3.1 The gravitational perturbation force due to a planet on the satellite-Earth two body system (Fig. 3.1) is

$$\vec{F}_d = \vec{F}_{ps} - \vec{F}_p = Gmm_p\left(\frac{\vec{r}_{ps}}{r_{ps}^3} - \frac{\vec{r}_p}{r_p^3}\right) \qquad (3.1)$$

where m is the mass of the satellite, m_p is the mass of the planet, and G is the universal gravitational constant.

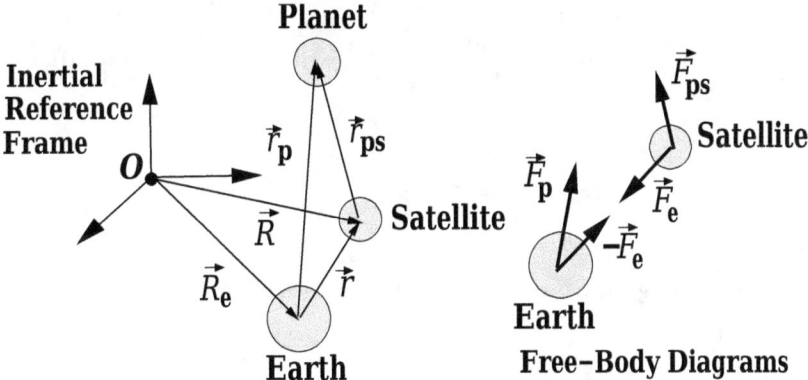

Figure 3.1: Planetary Gravitational Perturbation.

Knowing $\vec{r}_{ps} = \vec{r}_p - \vec{r}$, the gravitational perturbation acceleration $(\vec{F}_d/m)$

can be written as

$$\vec{f}_d = \mu_p \left(\frac{\vec{r}_p - \vec{r}}{|\vec{r}_p - \vec{r}|^3} - \frac{\vec{r}_p}{r_p^3} \right)$$

$$= \mu_p \left(\frac{\vec{r}_p - \vec{r}}{[r_p^2 - 2(\vec{r}_p \cdot \vec{r}) + r^2]^{3/2}} - \frac{\vec{r}_p}{r_p^3} \right) \tag{3.2}$$

where $\mu_p = Gm_p$ is the gravitational parameter of the planetary body. Considering the fact $r \ll r_p$, we can approximate $\vec{f}_d$ by carrying out Binomial series expansion till $\mathcal{O}(1/r_p^2)$ as

$$\vec{f}_d = \mu_p \left\{ \frac{\vec{r}_p - \vec{r}}{r_p^3} \left[1 - 2\frac{(\vec{r}_p \cdot \vec{r})}{r_p^2} + \frac{r^2}{r_p^2} \right]^{-3/2} - \frac{\vec{r}_p}{r_p^3} \right\}$$

$$= \frac{\mu_p}{r_p^3} \left\{ (\vec{r}_p - \vec{r}) \left[1 + 3\frac{(\vec{r}_p \cdot \vec{r})}{r_p^2} + \mathcal{O}(\frac{1}{r_p^2}) \right] - \vec{r}_p \right\} \tag{3.3}$$

or

$$\vec{f}_d = \frac{\mu_p}{r_p^3} \left\{ -\vec{r} + 3\frac{(\vec{r}_p \cdot \vec{r})}{r_p^2}\vec{r}_p \right\} \tag{3.4}$$

Writing $\vec{r}$ and $\vec{r}_p$ in terms of unit vectors ($\vec{r} = r\hat{i}$ and $\vec{r}_p = r_p\hat{i}_p$) leads to

$$\vec{f}_d = \frac{\mu_p}{r_p^3}r \left[3(\hat{i} \cdot \hat{i}_p)\hat{i}_p - \hat{i} \right] \tag{3.5}$$

We express the perturbation force $\vec{f}_d$ along orbital reference frame $\hat{i}\hat{j}\hat{k}$ as

$$f_x = \vec{f}_p \cdot \hat{i} = \frac{\mu_p r}{r_p^3} \left[3(\hat{i} \cdot \hat{i}_p)(\hat{i}_p \cdot \hat{i}) - 1 \right] \tag{3.6}$$

$$f_y = \vec{f}_p \cdot \hat{j} = \frac{3\mu_p r}{r_p^3}(\hat{i} \cdot \hat{i}_p)(\hat{i}_p \cdot \hat{j}) \tag{3.7}$$

$$f_z = \vec{f}_p \cdot \hat{k} = \frac{3\mu_p r}{r_p^3}(\hat{i} \cdot \hat{i}_p)(\hat{i}_p \cdot \hat{k}) \tag{3.8}$$

3.2 Aerodynamic drag on a satellite assuming it is a point mass (m), is given by

$$\vec{F}_d = -\frac{1}{2}C_D \rho A V_{rel}^2 \frac{\vec{V}_{rel}}{V_{rel}} \tag{3.9}$$

where C_D is the drag coefficient, A is the area of the satellite surface perpendicular to $\vec{V}_{rel}$, V_{rel} is the velocity of the satellite relative to the atmosphere, and ρ is the density of the atmosphere.

Assuming the atmosphere is stationary, $\vec{V}_{rel}$ equals to the satellite velocity $\vec{v}$. In order to find the effect of the aerodynamic drag on the satellite orbital parameters, we will use Lagrange planetary equations of motion. We are required to find the components of the aerodynamic drag along $\hat{i}$, $\hat{j}$, and $\hat{k}$ (orthogonal right-handed unit vectors along r (orbital radius), θ (true anomaly), and z directions). The velocity of the satellite $\vec{v}$ can be expressed as

$$\vec{v} = \dot{r}\hat{i} + r\dot{\theta}\hat{j} \tag{3.10}$$

Using the orbital motion relations $h = \sqrt{\mu p}$ and $r = p/(1 + e\cos\theta)$, we get

$$\dot{r} = \frac{he\sin\theta}{p} \tag{3.11}$$

$$\dot{\theta} = \frac{h}{r^2} \tag{3.12}$$

where h is the orbital angular momentum per unit mass of the satellite, e is the orbital eccentricity, and p is the semi-latus rectum.

Substituting the values of $\dot{r}$ and $\dot{\theta}$ from Eqs. (3.11-3.12) into Eq.(3.10) and applying the orbital motion relations, we get

$$v^2 = \dot{r}^2 + r^2\dot{\theta}^2 = \frac{\mu}{p}(1 + e^2 + 2e\cos\theta) \tag{3.13}$$

Using Eqs. (3.10-3.13), the aerodynamic drag as per Eq. (3.9) can be written along $\hat{i}$, $\hat{j}$, and $\hat{k}$ as follows:

$$f_x = -\frac{1}{2m}C_D\rho A v^2 \frac{e\sin\theta}{(1 + e^2 + 2e\cos\theta)^{1/2}} \tag{3.14}$$

$$f_y = -\frac{1}{2m}C_D\rho A v^2 \frac{1 + e\cos\theta}{(1 + e^2 + 2e\cos\theta)^{1/2}} \tag{3.15}$$

$$f_z = 0 \tag{3.16}$$

3.4 The solar radiation force on a highly reflective surface (*i.e.*, $\rho_a \approx 0$ and $\rho_d \approx 0$) is

$$\begin{aligned}\vec{F} &= 2\rho_s p A H(\cos\zeta)(\hat{s} \cdot \hat{n})^2 \hat{n} = 2\rho_s p A |\hat{s} \cdot \hat{n}|(\hat{s} \cdot \hat{n})\hat{n} \\ &= 2\rho_s p A |\cos\zeta|\cos\zeta\hat{n}\end{aligned} \tag{3.17}$$

Here $(\hat{s} \cdot \hat{n}) = \cos\zeta$. The expression for $\hat{s}$ in terms of $\hat{I}_n$-$\hat{J}_n$-$\hat{K}_n$ coordinate frame (where $\hat{I}_n$ in the direction towards the ascending node; $\hat{K}_n$ is perpendicular to the orbit plane along $\vec{h}$, and $\hat{K}_n \times \hat{I}_n = \hat{J}_n$) as

$$\hat{s} = -\cos(\psi - \Omega)\hat{I}_n - \sin(\psi - \Omega)\cos(i - \epsilon)\hat{J}_n + \sin(\psi - \Omega)\sin(i - \epsilon)\hat{K}_n \tag{3.18}$$

where ψ is the Sun angle with respect to the Vernal equinox, and ϵ is the angle between the equatorial and the ecliptic plane ($\epsilon = \pm$ 23 deg 27 min). The $\hat{I}_n$, $\hat{J}_n$, and $\hat{K}_n$ vectors can be expressed in terms of the satellite coordinates i, j and k as

$$\left\{ \begin{array}{c} \hat{I}_n \\ \hat{J}_n \\ \hat{K}_n \end{array} \right\} = \left[\begin{array}{ccc} cos(\omega + \theta) & -sin(\omega + \theta) & 0 \\ sin(\omega + \theta) & cos(\omega + \theta) & 0 \\ 0 & 0 & 1 \end{array} \right] \left\{ \begin{array}{c} \hat{i} \\ \hat{j} \\ \hat{k} \end{array} \right\} \qquad (3.19)$$

The unit vector $\hat{n}$ is with respect to the satellite body fixed frame.

3.6 The Earth's magnetic field with respect to the orbital reference frame is

$$\vec{B} = \frac{\mu_f}{R^3} \left\{ \begin{array}{c} cosi_m \\ -2sin(\omega + \theta)sini_m \\ cos(\omega + \theta)sini_m \end{array} \right\} \qquad (3.20)$$

For a satellite in equatorial orbit ($i.e., \hat{i}_m \approx 0$), the Earth's magnetic field simplifies to

$$\vec{B} = \frac{\mu_f}{R^3} \left\{ \begin{array}{c} 1 \\ 0 \\ 0 \end{array} \right\} \qquad (3.21)$$

or

$$B = \frac{\mu_f}{R^3} \qquad (3.22)$$

For a satellite in polar orbit ($i.e., \hat{i}_m \approx 0$), the Earth's magnetic field reduces to

$$\vec{B} = \frac{\mu_f}{R^3} \left\{ \begin{array}{c} 0 \\ -2sin(\omega + \theta) \\ cos(\omega + \theta) \end{array} \right\} = \frac{\mu_f}{R^3}\hat{i}_o \qquad (3.23)$$

or

$$B = \frac{\mu_f}{R^3} \sqrt{[4sin^2(\omega + \theta) + cos^2(\omega + \theta)]} \qquad (3.24)$$

Approximating the term inside the square root by taking the average,

$$\frac{1}{2\pi} \int_0^{2\pi} \left[4sin^2(\omega + \theta) + cos^2(\omega + \theta) \right] d(\omega + \theta) = \frac{5}{2} \qquad (3.25)$$

leads to the Earth's magnetic field as

$$B = \frac{1.581\mu_f}{R^3} \approx \frac{2\mu_f}{R^3} \qquad (3.26)$$

3.7 The Earth's magnetic field is

$$\vec{B} = \frac{\mu_f}{R^3}[3(\hat{i}_m \cdot \hat{i}_R)\hat{i}_R - \hat{i}_m] \qquad (3.27)$$

For a satellite in a circular orbit, we have

$$\hat{k}_o = \hat{i}_R \times \hat{i}_V \qquad (3.28)$$

where $\hat{i}_R$ and $\hat{i}_V$ are unit vectors along the satellite position and velocity vectors $\vec{R}$ and $\vec{V}$, respectively.

The the components of $\vec{B}$ along the pitch z-axis can be expressed as

$$B_z = \vec{B} \cdot \hat{k}_o = \frac{\mu_f}{R^3}[3(\hat{i}_m \cdot \hat{i}_R)\hat{i}_R \cdot (\hat{i}_R \times \hat{i}_V) - \hat{i}_m \cdot (\hat{i}_R \times \hat{i}_V)] \quad (3.29)$$

Applying the property of triple product of three vectors, $\vec{a}.(\vec{b}\times\vec{c})=\vec{a}.(\vec{b}\times\vec{c})$, we can write

$$\hat{i}_R \cdot (\hat{i}_R \times \hat{i}_V) = (\hat{i}_R \times \hat{i}_R) \cdot \hat{i}_V = 0 \qquad (3.30)$$

Substituting this in Eq. (3.29), we get

$$B_z = -\frac{\mu_f}{R^3}[\hat{i}_m \cdot (\hat{i}_R \times \hat{i}_V)] \qquad (3.31)$$

Since the angular momentum of the orbit given by $\vec{h} = \vec{R} \times \vec{V} = RV(\hat{i}_R\times\hat{i}_V)$ is constant, and the direction of the magnetic dipole vector, $\hat{i}_m$ is nearly constant, B_z is nearly constant. Thus, the components of $\vec{B}$ along the pitch axis remain constant when the satellite is in a circular orbit.

Chapter 4

Dynamics I

4.1 Given.

At time $t=0$,

$a=46800$ km
$e=0.85$
$\theta=52$ deg

Find.

$t - t_p$

Solution.

Consider *Kepler's time equation,*

$$M = n(t - t_p) = E - esinE$$

To find $t - t_p$ we need to determine an Eccentric anomaly E and mean angular velocity n as follows:

$$E = 2tan^{-1}\left[\sqrt{\frac{1-e}{1+e}}tan\frac{\theta}{2}\right]$$

$$= 2tan^{-1}\left[\sqrt{\frac{1-0.85}{1+0.85}}tan\frac{52deg}{2}\right] = 0.2759rad$$

$$n = \sqrt{\frac{\mu}{a^3}} = \sqrt{\frac{3.986 \times 10^5}{46800^3}} = 6.2359 \times 10^{-5} \text{rad/s}$$

Thus, the time that has elapsed since the satellite passed through periapsis can be obtained as

$$t - t_p = \frac{E - esinE}{n}$$

$$0 - t_p = \frac{0.2759 - 0.85sin(0.2759)}{6.2359 \times 10^{-5} \text{rad/s}} = 711.5sec$$

The spacecraft passed through periapsis 711.2 sec before t=0.

4.2

$$E = 2tan^{-1}\left[\sqrt{\frac{1-e}{1+e}}tan\frac{\theta}{2}\right]$$

$$= 2tan^{-1}\left[\sqrt{\frac{1-0.85}{1+0.85}}tan\frac{297deg}{2}\right] = -19.8deg$$

Since the spacecraft θ at a later time,

E=-19.8 deg + 360 deg=340.2 deg= 5.938 rad

Now apply Kepler's equation, this time with known t_0=-711.2 sec:

$$t = t_0 + \frac{5.938rad - 0.85sin(5.938rad)}{6.231 \times 10^{-5}sec^{-1}}$$

or

$$t = -711.2sec + \frac{5.938rad - 0.85sin(5.938rad)}{6.231 \times 10^{-5}sec^{-1}} = 99122sec$$

4.3

$$\phi = tan^{-1}\frac{v_r}{v_\theta}$$

$$= tan^1\left(\frac{-3.475}{5.940}\right) = -30.3deg$$

Note ϕ is negative because v_r is negative.

b) Orbital period

$$E = -\frac{\mu}{r} + \frac{v^2}{2} = -\frac{\mu}{r} + \frac{v_r^2 + v_\theta^2}{2}$$
$$= -7.969 \quad \text{km}^2/\text{s}^2$$

$$a = -\frac{\mu}{2E} = 25009.411 \text{ km}$$

$$T = \frac{2\pi a^{3/2}}{\mu} = 39361 \text{ sec}$$

c) eccentricity

The momentum is

$$\vec{h} = \vec{r} \times \vec{v}$$
$$= rv\sin(90^\circ - \phi)\hat{k} = rv\cos\phi\hat{k}$$
$$= rv_\theta\hat{k} = 74819.646 \text{ km}^2/s$$

$$p = \frac{h^2}{\mu} = 14044 \text{ km}$$

Using $p = a(1 - e^2)$, we solve for e as

$$e = \sqrt{1 - p/a} = 0.6621$$

d) true anomaly

$$r = \frac{p}{1 + e\cos\theta}$$

$$\cos\theta = \frac{1}{e}\left(\frac{p}{r} - 1\right) = 0.1736$$

$$\theta = \pm cos^{-1}(0.1736)$$

Use $-$ sign because $\phi < 0$, so

$$\theta = -80 \text{ deg}$$

4.4

$$\mathcal{E} = \frac{v_0^2}{2} - \frac{\mu}{r_0} = \frac{-\mu}{2a}$$

$$h = r_0 v_0 cos\beta_0$$

$$p = \frac{h^2}{\mu} = \frac{(r_0 v_0 cos\beta_0)^2}{\mu}$$

$$a(1 - e^2) = \frac{h^2}{\mu} = \frac{(r_0 v_0 cos\beta_0)^2}{\mu}$$

$$e = \sqrt{1 + \frac{(r_0 v_0 cos\beta_0)^2(v_0^2 - 2\mu/r_0)}{\mu^2}}$$

4.5 The orbital position is given by

$$\vec{r} = rcos\theta\hat{i}_e + rsin\theta\hat{i}_p$$

Differentiating with respect to time and considering $d\hat{i}_e/dt = d\hat{i}_p/dt = 0$), we have

$$\dot{\vec{r}} = [\dot{r}cos\theta + rt\dot{a}sin\theta]\hat{i}_e + [\dot{r}sin\theta + r\dot{\theta}cos\theta]\hat{i}_p$$

Here r, $\dot{r}$, and $\dot{\theta}$ are obtained as described next.

$$r = \frac{p}{1 + ecos\theta}$$

$$\vec{h} = \vec{r} \times \vec{v} = r^2\dot{\theta}$$

$$\dot{r} = \frac{dr}{dt} = \frac{d}{dt}\left(\frac{p}{1+ecos\theta}\right) = \frac{pe\dot{\theta}sin\theta}{(1+ecos\theta)^2}$$

$$= \frac{r^2 e\dot{\theta}sin\theta}{p} = \frac{h}{p}esin\theta = \frac{\mu}{h}esin\theta$$

$$\dot{\theta} = \frac{h}{r^2} = \frac{h(1+ecos\theta)^2}{p^2} = \frac{\mu^2(1+ecos\theta)^2}{h^3}$$

$$\vec{v} = -\frac{\mu}{h}sin\theta\hat{i}_e + \frac{\mu}{h}(e+cos\theta)\hat{i}_p$$

4.6 *Part I.*

The J_2-perturbation affects (on average) the *Longitude of the ascending node*, Ω, the *Argument of the periapsis*, ω, and the *Mean anomaly*. The effects on orbital elements are as follows:

Constant (on average): a, e, i

Changing (on average): ω, Ω, M

Part II.

Given.

a=7000 km
e=0.08
i=28.5 deg

Find.

$\left(\frac{d\Omega}{dt}\right)_{avg}$

Solution.

The average rate of change of the longitude of the ascending node is

$$\left(\frac{d\Omega}{dt}\right)_{avg} = \frac{-3J_2R_e^2n}{2p^2}cosi$$

Using the relations $n = \sqrt{\mu/a^3}$ and $p = a(1-e^2)$, and substituting J_2=1.0826× 10^{-3}, R_e=6378 km, and $\mu = 3.986 \times 10^5$ km^3/s^2, we have

$$\left(\frac{d\Omega}{dt}\right)_{avg} = \frac{-3J_2R_e^2\mu^{1/2}cosi}{2a^{7/2}(1-e^2)^2}$$

$$= \frac{-3(1.0826 \times 10^{-3})(6378 \text{ km})^2(3.986 \times 10^5 \text{ km}^3/\text{s}^2)^{1/2}cos(28.5^o)}{2(7000 \text{ km})^{7/2}[1-(0.08)^2]^2}$$

$$= -1.294 \times 10^{-6} \text{ rad/sec} = -6.4^o/\text{day}$$

Chapter 5

Dynamics II

Problem Set 5

5.1 Newton Method

The attitude equation of motion of the dumbbell system is given by

$$\vec{T} = \dot{\vec{H}} \tag{5.1}$$

where the attitude angular momentum of the system $\vec{H}$ is

$$\vec{H} = M_e L^2 \left[\dot{\eta}\hat{j} + (\dot{\theta} + \dot{\beta})cos\eta\hat{k} \right] \tag{5.2}$$

Note, the above expression can be easily derived using the rigid body analogy where the attitude angular momentum of the rigid satellite about the principal moments of inertia axes is given by

$$\vec{H} = I_x \omega_x \hat{i} + I_y \omega_y \hat{j} + I_z \omega_z \hat{k} \tag{5.3}$$

For the dumbbell system, $I_x=0$ and $I_y=I_z=M_e L^2$ with $M_e=(m_1 m_2)/(m_1+m_2)$. The angular velocity vector is

$$\vec{\omega} = \omega_x \hat{i} + \omega_y \hat{j} + \omega_z \hat{k} \tag{5.4}$$

where

$$\omega_x = -(\dot{\theta} + \dot{\beta})sin\eta, \quad \omega_y = \dot{\eta}, \quad \omega_z = (\dot{\theta} + \dot{\beta})cos\eta$$

Substituting I_k, $k = x, y, z$ and ω_k, $k = x, y, z$=0 into Eq. (5.3) result in the angular momentum of the dumbbell system given by Eq. (5.2).

Differentiating Eq. (5.2) with respect to time and assuming the system is in a circular orbit ($i.e., \ddot{\theta} = 0$) yield the rate of change of the system angular momentum as

$$
\begin{aligned}
\dot{\vec{H}} &= \dot{\vec{H}}_{xyz} + \vec{\omega} \times \vec{H} \\
&= M_e L^2 \left\{ \ddot{\eta}\hat{j} + \left[(\ddot{\theta} + \ddot{\beta})cos\eta - (\dot{\theta} + \dot{\beta})\dot{\eta}sin\eta \right] \hat{k} \right\} \\
&\quad + M_e L^2 \vec{\omega} \times \left[\dot{\eta}\hat{j} + +(\ddot{\theta} + \ddot{\beta})cos\eta\hat{k} \right] \\
&= M_e L^2 \left[\ddot{\eta} + (\dot{\theta} + \dot{\beta})^2 sin\eta cos\eta \right] \hat{j} + M_e L^2 [\ddot{\beta}cos\eta - 2(\dot{\theta} + \dot{\beta})\dot{\eta}sin\eta]\hat{k}
\end{aligned}
\tag{5.5}
$$

Next we derive the external toque $\vec{T}$ due to gravitational force. The gravitational force exerted on mass m_1 at a distance of R_1 from the center of Earth is given by

$$
\vec{F}_1 = -\frac{\mu m_1 \vec{R}_1}{R_1^3}
\tag{5.6}
$$

Thus, the torque exerted on a mass of m_1 is

$$
\vec{T}_1 = m_1 \vec{r}_1 \times \vec{F} = -\mu m_1 \vec{r}_1 \times \frac{\vec{R}_1}{R_1^3}
\tag{5.7}
$$

Here $\vec{R}_1 = \vec{R} + \vec{r}_1$. Substituting $|\vec{R}_1|$ in Eq.(5.7), we get

$$
\begin{aligned}
\vec{T}_1 &= -\mu m_1 \vec{r}_1 \times \frac{(\vec{R} + \vec{r}_1)}{(R^2 + 2\vec{R} \cdot \vec{r}_1 + r_1^2)^{3/2}} \\
&= -\frac{\mu m_1}{R^3}(\vec{r}_1 \times \vec{R}) \left[1 + \frac{2\vec{R} \cdot \vec{r}_1}{R^2} + \frac{r_1^2}{R^2} \right]^{-3/2}
\end{aligned}
\tag{5.8}
$$

Applying Binomial series expansion for the term inside the bracket, we get

$$
\vec{T}_1 = -\frac{\mu m_1}{R^3}(\vec{r}_1 \times \vec{R}) \left[1 - \frac{3(\vec{R} \cdot \vec{r}_1)}{R^2} - \frac{3}{2}\frac{r_1^2}{R^2} + \frac{15}{2}\frac{(\vec{R} \cdot \vec{r}_1)^2}{R^4} + \cdots \right]
\tag{5.9}
$$

Now, considering the fact $r_1 \ll R$ and carrying out expansion untill

$\mathcal{O}(1/R^3)$, we get

$$\vec{T_1} = -\frac{\mu m_1}{R^3}(\vec{r_1} \times \vec{R})\left[1 - \frac{3(\vec{R}\cdot\vec{r_1})}{R^2}\right] + \mathcal{O}(\frac{1}{R^3})$$

$$= \frac{\mu\vec{R}}{R^3} \times \left[m_1\vec{r_1} - m_1\frac{3(\vec{R}\cdot\vec{r_1})\vec{r_1}}{R^2}\right] \tag{5.10}$$

Similarly, the torque exerted on mass m_2 can be obtained as

$$\vec{T_2} = \frac{\mu\vec{R}}{R^3} \times \left[m_2\vec{r_2} - m_2\frac{3(\vec{R}\cdot\vec{r_2})\vec{r_2}}{R^2}\right] \tag{5.11}$$

Thus, the total torque exerted on the system is the sum of the torques exerted on mass m_1 and m_2. Adding T_1 and T_2, we obtain the total torque

$$\vec{T} = \frac{\mu\vec{R}}{R^3} \times \left\{m_1\vec{r_1} + m_2\vec{r_2} - \frac{3}{R^2}[m_1(\vec{R}\cdot\vec{r_1})\vec{r_1} + m_2(\vec{R}\cdot\vec{r_2})\vec{r_2}]\right\} \tag{5.12}$$

Knowing

$$\vec{r_1} = -\frac{m_2}{m_1 + m_2}\vec{L}$$

$$\vec{r_2} = \frac{m_1}{m_1 + m_2}\vec{L}$$

and

$$m_1\vec{r_1} + m_2\vec{r_2} = 0$$

the preceding equation simplifies to

$$\vec{T} = -\frac{3\mu}{R^3}M_eL^2(\hat{i}_o\cdot\hat{i})(\hat{i}_o\times\hat{i}) \tag{5.13}$$

Knowing

$$\hat{i}_o = cos\beta cos\eta\hat{i} - sin\beta\hat{j} + cos\beta sin\eta\hat{k}$$

we have

$$\vec{T} = -\frac{3\mu}{R^3}M_eL^2\left[sin\eta cos\eta cos^2\beta\hat{j} + sin\beta cos\beta cos\eta\hat{k}\right] \tag{5.14}$$

Using the preceding equation and Eqs. (5.1) and (5.5) as well as rearranging the terms, we obtain equations of motion of the system as

$$\ddot{\beta} - 2(\dot{\theta} + \dot{\beta})\dot{\eta}tan\eta + \frac{3}{2}\dot{\theta}^2 sin2\beta = 0 \tag{5.15}$$

$$\ddot{\eta} + \frac{1}{2}(\dot{\theta} + \dot{\beta})^2 sin2\eta + \frac{3}{2}\dot{\theta}^2 sin2\eta cos^2\beta = 0 \tag{5.16}$$

where $\dot{\theta} = \sqrt{\mu/R^3}$.

Lagrange Method

The equations of motion of the system with generalized coordinates $q_1 = \beta$ and $q_2 = \eta$ are given by

$$\frac{d}{dt}\left(\frac{\partial T}{\partial \dot{q}_k}\right) - \frac{\partial T}{\partial q_k} + \frac{\partial U}{\partial q_k} = 0, \quad k = 1, 2 \tag{5.17}$$

The potential and kinetic energies of the system are

$$U = -\frac{\mu M}{R} + \frac{\mu}{2R^3}M_e(1 - 3cos^2\beta cos^2\eta)L^2 \tag{5.18}$$

$$T = \frac{1}{2}M(\dot{R}^2 + \dot{\theta}^2 R^2) + \frac{1}{2}M_e[(\dot{\theta} + \dot{\beta})^2 cos^2\eta + \dot{\eta}^2]L^2 \tag{5.19}$$

where $M = m_1 + m_2$ and $M_e = m_1 m_2/(m_1 + m_2)$.

Using the preceding relations, the equations of motion of the system are derived as follows:

β-equation

$$\frac{d}{dt}\left(\frac{\partial T}{\partial \dot{\beta}}\right) - \frac{\partial T}{\partial \beta} + \frac{\partial U}{\partial \beta} = 0 \tag{5.20}$$

Here

$$\frac{\partial T}{\partial \dot{\beta}} = M_e L^2(\dot{\theta} + \dot{\beta})cos^2\eta \tag{5.21}$$

$$\Rightarrow \frac{d}{dt}\left(\frac{\partial T}{\partial \dot{\beta}}\right) = M_e L^2[\ddot{\beta}cos^2\eta - 2(\dot{\theta} + \dot{\beta})cos\eta sin\eta] \tag{5.22}$$

$$\frac{\partial T}{\partial \beta} = 0 \tag{5.23}$$

$$\frac{\partial U}{\partial \beta} = \frac{3\mu}{R^3}M_e L^2 cos\beta sin\beta cos^2\eta \tag{5.24}$$

Thus, β-equation of motion is

$$M_e L^2[\ddot{\beta}cos^2\eta - 2(\dot{\theta} + \dot{\beta})\dot{\eta}cos\eta sin\eta] + \frac{3\mu}{R^3}M_e L^2 cos\beta sin\beta cos^2\eta = 0 \tag{5.25}$$

η-equation

$$\frac{d}{dt}\left(\frac{\partial T}{\partial \dot{\eta}}\right) - \frac{\partial T}{\partial \eta} + \frac{\partial U}{\partial \eta} = 0 \tag{5.26}$$

Here

$$\frac{\partial T}{\partial \dot{\eta}} = M_e \dot{\eta} L^2 \qquad (5.27)$$

$$\Rightarrow \frac{d}{dt}\left(\frac{\partial T}{\partial \dot{\eta}}\right) = M_e \ddot{\eta} L^2 \qquad (5.28)$$

$$\frac{\partial T}{\partial \eta} = -M_e L^2(\dot{\theta} + \dot{\beta})^2 cos\eta sin\eta \qquad (5.29)$$

$$\frac{\partial U}{\partial \eta} = \frac{3\mu}{R^3} M_e L^2 cos\eta sin\eta cos^2\beta \qquad (5.30)$$

Thus, η-equation of motion is

$$M_e L^2 \ddot{\eta} + M_e L^2(\dot{\theta} + \dot{\beta})^2 cos\eta sin\eta + \frac{3\mu}{R^3} M_e L^2 cos\eta sin\eta cos^2\beta = 0 \qquad (5.31)$$

Thus, the equations of motion of the system can be rewritten as

$$\ddot{\beta} - 2(\dot{\theta} + \dot{\beta})\dot{\eta} tan\eta + \frac{3}{2}\dot{\theta}^2 sin2\beta = 0 \qquad (5.32)$$

$$\ddot{\eta} + \frac{1}{2}(\dot{\theta} + \dot{\beta})^2 sin2\eta + \frac{3}{2}\dot{\theta}^2 sin2\eta cos^2\beta = 0 \qquad (5.33)$$

where $\dot{\theta} = \sqrt{\mu/R^3}$.

Remarks. 1. In the above derivations leading to the attitude equations of motion of the system assume that the orbital motion of the system does not affect the system attitude motion.

2. In this example problem we have derived three-dimensional attitude equations of motion of the system using the Euler's method as well as the Lagrange method. However, these equations of motion are different before final simplifications. These can be explained by deriving generalized forces Q_k, $k = \beta, \eta$ using *virtual work method*. The virtual angular displacement λ of the system is given by

$$\vec{\lambda} = \beta\hat{k}_o + \eta\hat{j} \qquad (5.34)$$

The generalized forces Q_k, $k = \beta, \eta$ are given by

$$Q_\beta = \vec{T} \cdot \frac{\partial\vec{\lambda}}{\partial\beta}, \quad Q_\eta = \vec{T} \cdot \frac{\partial\vec{\lambda}}{\partial\eta} \qquad (5.35)$$

Using the relation (5.34) yields

$$Q_\beta = \vec{T} \cdot \hat{k}_o, \quad Q_\eta = \vec{T} \cdot \hat{j} \qquad (5.36)$$

Knowing

$$\vec{T} = T_\eta \hat{j} + T_\beta \hat{k}$$

and

$$\hat{k}_o = -sin\eta\hat{i} + cos\eta\hat{k}$$

we have

$$Q_\beta = T_\beta cos\eta \qquad (5.37)$$
$$Q_\eta = T_\eta \qquad (5.38)$$

Thus, we can write the Lagrange equations of motion $(L_k,\ k = \beta, \eta)$ which are related to the Euler's equations of motion $(E_k,\ k = \beta, \eta)$ as

$$L_\beta = E_\beta cos\eta \qquad (5.39)$$
$$L_\eta = E_\eta \qquad (5.40)$$

5.2 The kinetic and potential energies of the system are

$$T = \frac{1}{2}M\dot{R}^2 + \frac{1}{2}M_{t1}\omega_1^2 L_1^2 + \frac{1}{2}M_{t2}\omega_2^2 L_2^2 + M_{t3}\omega_1\omega_2 L_1 L_2 cos(\beta_1 - \beta_2)$$

$$(5.41)$$

$$U = -\frac{\mu M}{R} + \frac{\mu}{2R^3}\left\{ M_{t1}L_1^2 + M_{t2}L_2^2 + 2M_{t3}L_1 L_2 cos(\beta_1 - \beta_2) \right\}$$

$$- \frac{3\mu}{2R^3}\left\{ M_{t1}L_1^2 cos^2\beta_1 + M_{t2}L_2^2 cos^2\beta_2 + 2M_{t3}L_1 L_2 cos\beta_1 cos\beta_2 \right\}$$

$$(5.42)$$

where $M = m_1 + m_2 + m_3$, $\omega_1 = \dot{\theta} + \dot{\beta}_1$, $\omega_2 = \dot{\theta} + \dot{\beta}_2$, and

$$M_{t1} = m_1\gamma_1^2 + m_2(1 - \gamma_1)^2 + m_3(1 - \gamma_1)^2$$
$$M_{t2} = m_1\gamma_2^2 + m_2\gamma_2^2 + m_3(1 - \gamma_2)^2$$
$$M_{t3} = m_1\gamma_1\gamma_2 - m_2(1 - \gamma_1)\gamma_2 + m_3(1 - \gamma_1)(1 - \gamma_2)$$

with $\gamma_1 = (m_2 + m_3)/M$ and $\gamma_2 = m_3/M$.

Using the above relations for T and U, the equations of motion of the system are derived as follows:

β_1-**equation**

$$\frac{d}{dt}\left(\frac{\partial T}{\partial \dot{\beta}_1} \right) - \frac{\partial T}{\partial \beta_1} + \frac{\partial U}{\partial \beta_1} = 0 \qquad (5.43)$$

Here

$$\frac{\partial T}{\partial \dot{\beta_1}} = M_{t1}\omega_1 L_1^2 + M_{t3}\omega_2 L_1 L_2 cos(\beta_1 - \beta_2)$$

$$\Rightarrow \frac{d}{dt}\left(\frac{\partial T}{\partial \dot{\beta_1}}\right) = M_{t1}\dot{\omega}_1 L_1^2 + M_{t3}\dot{\omega}_2 L_1 L_2 cos(\beta_1 - \beta_2)$$

$$- M_{t3}\omega_2 L_1 L_2 (\dot{\beta}_1 - \dot{\beta}_2)sin(\beta_1 - \beta_2) \qquad (5.44)$$

$$\frac{\partial T}{\partial \beta_1} = - M_{t3}\omega_1\omega_2 L_1 L_2 sin(\beta_1 - \beta_2) \qquad (5.45)$$

$$\frac{\partial U}{\partial \beta_1} = - \frac{\mu}{R^3} M_{t3} L_1 L_2 sin(\beta_1 - \beta_2)$$

$$+ \frac{3\mu}{R^3}\left\{ M_{t1} L_1^2 cos\beta_1 sin\beta_1 + M_{t3} L_1 L_2 sin\beta_1 cos\beta_2 \right\} \qquad (5.46)$$

The β_1-equation of motion is

$$M_{t1}\dot{\omega}_1 L_1^2 + M_{t3}\dot{\omega}_2 L_1 L_2 cos(\beta_1 - \beta_2) + M_{t3}\omega_2 L_1 L_2 sin(\beta_1 - \beta_2)\Big[-(\dot{\beta}_1 - \dot{\beta}_2)$$

$$+ \omega_1 \Big] - \frac{\mu}{R^3} M_{t3} L_1 L_2 sin(\beta_1 - \beta_2) + \frac{3\mu}{R^3}\left\{ M_{t1} L_1^2 cos\beta_1 sin\beta_1 \right.$$

$$\left. + M_{t3} L_1 L_2 sin\beta_1 cos\beta_2 \right\} = 0 \qquad (5.47)$$

Dividing by $M_{t1}L_1^2$ and taking $M_{p1} = M_{t3}/M_{t1} = m_3/(m_2 + m_3)$ and knowing

$$\omega_k = \dot{\theta} + \dot{\beta}_k, \quad k = 1, 2 \dot{\omega}_k = \quad \ddot{\theta} + \ddot{\beta}_k = \ddot{\beta}_k, \quad k = 1, 2 \text{ as } \ddot{\theta} = 0 \text{ for circular orbit}$$

we have

$$\ddot{\beta}_1 + M_{p1}\left(\frac{L_2}{L_1}\right)cos(\beta_1 - \beta_2)\ddot{\beta}_2 + M_{p1}\left(\frac{L_2}{L_1}\right)sin(\beta_1 - \beta_2)$$

$$\times \left\{ -\omega_2(\dot{\beta}_1 - \dot{\beta}_2) + \omega_1\omega_2 \right\} - \frac{\mu}{R^3} M_{p1}\left(\frac{L_2}{L_1}\right)sin(\beta_1 - \beta_2)$$

$$+ \frac{3\mu}{R^3}\left\{ cos\beta_1 sin\beta_1 + M_{p1}\left(\frac{L_2}{L_1}\right)sin\beta_1 cos\beta_2 \right\} = 0 \qquad (5.48)$$

Knowing $\mu/R^3 = \dot{\theta}^2$ and writing the derivative with respect to θ (i.e., $()' = d()/d\theta; ()'' = d^2()/d\theta^2$. Here θ denotes an angle with respect to a reference line), the preceding equation can be written as

$$\beta_1'' + M_{p1}\left(\frac{L_2}{L_1}\right)cos(\beta_1 - \beta_2)\beta_2'' + M_{p1}\left(\frac{L_2}{L_1}\right)\left\{\beta_2'(2 + \beta_2')sin(\beta_1 - \beta_2)\right.$$

$$\left. + 3sin\beta_1 cos\beta_2\right\} + 3sin\beta_1 cos\beta_1 = 0$$

$$(5.49)$$

β_2-equation

$$\frac{d}{dt}\left(\frac{\partial T}{\partial \dot{\beta}_2}\right) - \frac{\partial T}{\partial \beta_2} + \frac{\partial U}{\partial \beta_2} = 0 \qquad (5.50)$$

Here

$$\frac{\partial T}{\partial \dot{\beta}_2} = M_{t2}\omega_2 L_2^2 + M_{t3}\omega_1 L_1 L_2 cos(\beta_1 - \beta_2)$$

$$\Rightarrow \frac{d}{dt}\left(\frac{\partial T}{\partial \dot{\beta}_2}\right) = M_{t2}\dot{\omega}_2 L_2^2 + M_{t3}\dot{\omega}_1 L_1 L_2 cos(\beta_1 - \beta_2)$$

$$- M_{t3}\omega_1 L_1 L_2(\dot{\beta}_2 - \dot{\beta}_1)sin(\beta_2 - \beta_1) \qquad (5.51)$$

$$\frac{\partial T}{\partial \beta_2} = - M_{t3}\omega_1\omega_2 L_1 L_2 sin(\beta_2 - \beta_1) \qquad (5.52)$$

$$\frac{\partial U}{\partial \beta_2} = - \frac{\mu}{R^3}M_{t3}L_1 L_2 sin(\beta_2 - \beta_1)$$

$$+ \frac{3\mu}{R^3}\left\{M_{t2}L_2^2 cos\beta_2 sin\beta_2 + M_{t3}L_1 L_2 sin\beta_2 cos\beta_1\right\} \qquad (5.53)$$

The β_2-equation of motion is

$$M_{t2}\dot{\omega}_2 L_2^2 + M_{t3}\dot{\omega}_1 L_1 L_2 cos(\beta_1 - \beta_2) + M_{t3}\omega_1 L_1 L_2 sin(\beta_2 - \beta_1)\left[-(\dot{\beta}_2 - \dot{\beta}_1)\right.$$

$$\left. + \omega_2\right] - \frac{\mu}{R^3}M_{t3}L_1 L_2 sin(\beta_2 - \beta_1) + \frac{3\mu}{R^3}\left\{M_{t2}L_2^2 cos\beta_2 sin\beta_2\right.$$

$$\left. + M_{t3}L_1 L_2 sin\beta_2 cos\beta_1\right\} = 0 \qquad (5.54)$$

Dividing by $M_{t2}L_2^2$ and taking $M_{p2} = M_{t3}/M_{t2} = m_1/(m_1 + m_2)$ and knowing

$$\omega_k = \dot{\theta} + \dot{\beta}_k, \quad k = 1, 2\dot{\omega}_k = \quad \ddot{\theta} + \ddot{\beta}_k = \ddot{\beta}_k, \quad k = 1, 2 \text{ as } \ddot{\theta} = 0 \text{ for circular orbit}$$

we have

$$\ddot{\beta}_2 + M_{p2}\left(\frac{L_1}{L_2}\right)cos(\beta_1 - \beta_2)\ddot{\beta}_1 + M_{p2}\left(\frac{L_1}{L_2}\right)sin(\beta_2 - \beta_1)$$

$$\times \left\{ -\omega_1(\dot{\beta}_2 - \dot{\beta}_1) + \omega_1\omega_2 \right\} - \frac{\mu}{R^3}M_{p2}\left(\frac{L_1}{L_2}\right)sin(\beta_2 - \beta_1)$$

$$+ \frac{3\mu}{R^3}\left\{ cos\beta_2 sin\beta_2 + M_{p2}\left(\frac{L_1}{L_2}\right)sin\beta_2 cos\beta_1 \right\} = 0 \qquad (5.55)$$

Knowing $\frac{\mu}{R^3} = \dot{\theta}^2$ and writing the derivative with respect to θ (*i.e.*, $()' = d()/d\theta$; $()'' = d^2()/d\theta^2$. Here θ denotes an angle with respect to a reference line), the preceding equation can be written as

$$\beta_2'' + M_{p2}\left(\frac{L_1}{L_2}\right)cos(\beta_2 - \beta_1)\beta_1'' + M_{p2}\left(\frac{L_1}{L_2}\right)\left\{ \beta_1'(2 + \beta_1')sin(\beta_2 - \beta_1) \right.$$

$$\left. + 3sin\beta_2 cos\beta_1 \right\} + 3sin\beta_2 cos\beta_2 = 0$$

$$(5.56)$$

5. The system comprised of a satellite (m_1) and movable attached mass m_2 (see Fig. 5.1). The position of m_2 is defined with respect to the frame $S_1 - xyz$ as $\vec{L} = x\hat{i} + y\hat{j} + z\hat{k}$, where $\hat{i}$, $\hat{j}$, and $\hat{k}$ are unit vectors along the respective axes of the frame $S_1 - xyz$. Determine the angular momentum of the system orbiting in a circular orbit. Express the answer with respect to the frame $S_1 - xyz$.

Solution.

The given system is comprised of a rigid body (satellite; $M = 1$) and a point mass ($N = 1$). Referring to *Section 2.5: Summary (Kinematics: Rigid Body)*, the system angular momentum can be written as

$$\vec{H} = (m_1 + m_2)(\vec{R} \times \dot{\vec{R}}) + m_1(\vec{r}_1 \times \dot{\vec{r}}_1)$$

$$+ m_2(\vec{r}_2 \times \dot{\vec{r}}_2) + I_x\omega_x\hat{i} + I_y\omega_y\hat{j} + I_z\omega_z\hat{k} \qquad (5.57)$$

The first term corresponds to the orbital angular momentum H_o and it can be expressed as

$$\vec{H}_o = (m_1 + m_2)(\vec{R} \times \dot{\vec{R}}) = M[R\hat{i}_o \times (\dot{\theta}R\hat{j}_o)] = M\dot{\theta}^2 R\hat{k}_o \qquad (5.58)$$

where $M = m_1 + m_2$. The other terms in Eq. (5.57) correspond to the attitude angular momentum H_b and these can be expressed using the following relations:

$$\vec{r}_1 = -\gamma\vec{L}, \quad \vec{r}_2 = (1 - \gamma)\vec{L} \qquad (5.59)$$

$$\dot{\vec{r}}_1 = -\gamma(\dot{\vec{L}}_{xyz} + \vec{\omega} \times \vec{L}), \quad \dot{\vec{r}}_2 = (1 - \gamma)(\dot{\vec{L}}_{xyz} + \vec{\omega} \times \vec{L}) \qquad (5.60)$$

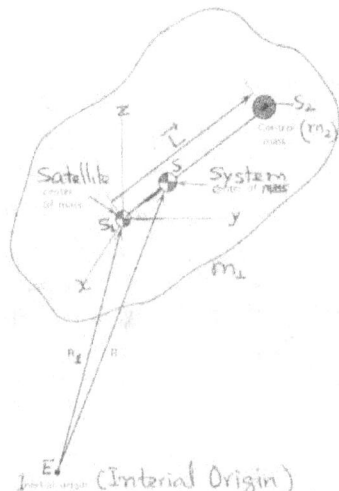

Figure 5.1: **Satellite (m_1) with attached mass (m_2).**

and

$$\vec{r}_1 \times \dot{\vec{r}}_1 = \gamma^2 (\vec{L} \times \dot{\vec{L}}_{xyz} + \vec{L} \times \vec{\omega} \times \vec{L}) \tag{5.61}$$

$$\vec{r}_2 \times \dot{\vec{r}}_2 = (1 - \gamma)^2 (\dot{\vec{L}}_{xyz} \times \vec{L} + \vec{\omega} \times \vec{L} \times \vec{L}) \tag{5.62}$$

Knowing $\vec{\omega} = \omega_x \hat{i} + \omega_y \hat{j} + \omega_z \hat{k}$ and $\vec{L} = x\hat{i} + y\hat{j} + z\hat{k}$, we have

$$\vec{L} \times \dot{\vec{L}}_{xyz} = (y\dot{z} + \dot{y}z)\hat{i} + (z\dot{x} - x\dot{z})\hat{j} + (x\dot{y} - y\dot{x})\hat{k} \tag{5.63}$$

$$\vec{L} \times \vec{\omega} \times \vec{L} = L^2\vec{\omega} - (\vec{L} \cdot \vec{\omega})\vec{L} \tag{5.64}$$

$$= L^2\vec{\omega} - (x\omega_x + y\omega_y + z\omega_z)\vec{L} \tag{5.65}$$

Thus, we can write the system angular momentum as

$$\vec{H} = \vec{H}_o + \vec{H}_b \tag{5.66}$$

where

$$\vec{H}_o = M\dot{\theta}^2 R\hat{k}_o, \quad \vec{H}_b = H_x\hat{i} + H_y\hat{j} + H_z\hat{k} \tag{5.67}$$

$$H_x = I_x\omega_x + M_e \left[(y^2 + z^2)\omega_x - (xy\omega_y + xz\omega_z) + y\dot{z} - \dot{y}z \right] \tag{5.68}$$

$$H_y = I_y\omega_y + M_e \left[(z^2 + x^2)\omega_y - (yz\omega_z + yx\omega_x) + z\dot{x} - \dot{x}z \right] \tag{5.69}$$

$$H_z = I_z\omega_z + M_e \left[(x^2 + y^2)\omega_z - (zx\omega_x + zy\omega_y) + x\dot{y} - \dot{x}y \right] \tag{5.70}$$

with

$$M_e = \frac{m_1 m_2}{m_1 + m_2} \tag{5.71}$$

5.1 Derive the 3-dimensional attitude equations of motion of a dumbbell system as described in Example 5.2 using Euler's method and Lagrange method.

Newton's Method

The translational and rotational equations of the motion are given by

$$\vec{F} = \frac{d\vec{p}}{dt} \tag{5.72}$$

$$\vec{T} = \frac{d\vec{H}}{dt} \tag{5.73}$$

where $\vec{F}$ and $\vec{T}$ denote total external force and torque acting at the center of mass of the system. The nomenclature $\vec{p}$ and $\vec{H}$ specify the system linear (orbital) momentum and angular (attitude) momentum, given by

$$\vec{p} = M\dot{\vec{R}}, \quad \vec{H} = M_e L^2 \left[\dot{\eta}\hat{j} + (\dot{\theta} + \dot{\beta})cos\eta\hat{k} \right] \tag{5.74}$$

where $M = m_1 + m_2$ and $M_e = m_1 m_2/(m_1 + m_2)$. Differentiating with respect to time and knowing

$$\ddot{\vec{R}} = \left(\ddot{R} - R\dot{\theta}^2 \right) \hat{i}_o + \left(R\ddot{\theta} + 2\dot{R}\dot{\theta} \right) \hat{j}_o$$

$$\vec{\omega} = [-(\dot{\theta} + \dot{\beta})sin\eta]\hat{i} + \dot{\eta}\hat{j} + [(\dot{\theta} + \dot{\beta})cos\eta]\hat{k}$$

we get

$$\frac{d\vec{p}}{dt} = M \left(\ddot{R} - R\dot{\theta}^2 \right) \hat{i}_o + M \left(R\ddot{\theta} + 2\dot{R}\dot{\theta} \right) \hat{j}_o \tag{5.75}$$

$$\frac{d\vec{H}}{dt} = M_e L^2 \left[\ddot{\eta} + (\dot{\theta} + \dot{\beta})^2 sin\eta cos\eta \right] \hat{j}$$

$$+ M_e L^2 [(\ddot{\theta} + \ddot{\beta})cos\eta - 2(\dot{\theta} + \dot{\beta})\dot{\eta}sin\eta]\hat{k} \tag{5.76}$$

Refer to the solution of Problem set 5.1 for the complete derivation of $\dot{\vec{H}}$.

Next we derive the external force and torque acting on the system. The external force vector is the sum of the gravitational forces acting on m_1 ($\vec{F_1}$) and m_2 ($\vec{F_2}$), given as

$$\vec{F} = \vec{F_1} + \vec{F_2} \tag{5.77}$$

where $\vec{F_1}$ is

$$\vec{F_1} = -\mu m_1 \frac{(\vec{R} + \vec{r_1})}{(R^2 + 2\vec{R} \cdot \vec{r_1} + r_1^2)^{3/2}}$$

$$= -\frac{\mu m_1}{R^3} (\vec{r_1} \times \vec{R}) \left[1 + \frac{2\vec{R} \cdot \vec{r_1}}{R^2} + \frac{r_1^2}{R^2} \right]^{-3/2} \tag{5.78}$$

Applying Binomial series expansion for the term inside the bracket, we get

$$\vec{F}_1 = -\frac{\mu m_1}{R^3}(\vec{R} + \vec{r}_1)\left[1 - \frac{3(\vec{R} \cdot \vec{r}_1)}{R^2} - \frac{3}{2}\frac{r_1^2}{R^2} + \frac{15}{2}\frac{(\vec{R} \cdot \vec{r}_1)^2}{R^4} + \cdots\right]$$

$$(5.79)$$

or

$$\vec{F}_1 = -\frac{\mu m_1}{R^3}\left\{\vec{R} + \vec{r}_1 - \frac{3(\vec{R} + \vec{r}_1)(\vec{R} \cdot \vec{r}_1)}{R^2} - \frac{3}{2}\frac{(\vec{R} + \vec{r}_1)r_1^2}{R^2}\right.$$
$$\left. + \frac{15}{2}\frac{(\vec{R} + \vec{r}_1)(\vec{R} \cdot \vec{r}_1)^2}{R^4} + \cdots\right\} \qquad (5.80)$$

Now, considering the fact $r_1 \ll R$ and carrying out expansion upto $\mathcal{O}(1/R^4)$ yields

$$\vec{F}_1 = -\frac{\mu m_1}{R^3}\left\{\vec{R} + \vec{r}_1 - \frac{3\vec{R}(\vec{R} \cdot \vec{r}_1)}{R^2} - \frac{3\vec{r}_1(\vec{R} \cdot \vec{r}_1)}{R^2} - \frac{3}{2}\frac{\vec{R}r_1^2}{R^2}\right.$$
$$\left. + \frac{15}{2}\frac{\vec{R}(\vec{R} \cdot \vec{r}_1)^2}{R^4}\right\} \qquad (5.81)$$

Similarly, the external force acting on mass m_2 is derived as

$$\vec{F}_2 = -\frac{\mu m_2}{R^3}\left\{\vec{R} + \vec{r}_2 - \frac{3\vec{R}(\vec{R} \cdot \vec{r}_2)}{R^2} - \frac{3\vec{r}_2(\vec{R} \cdot \vec{r}_2)}{R^2} - \frac{3}{2}\frac{\vec{R}r_2^2}{R^2}\right.$$
$$\left. + \frac{15}{2}\frac{\vec{R}(\vec{R} \cdot \vec{r}_2)^2}{R^4}\right\} \qquad (5.82)$$

Thus, the total external force acting on the system is

$$\vec{F} = -\frac{\mu}{R^3}\left\{(m_1 + m_2)\vec{R} + (m_1\vec{r}_1 + m_2\vec{r}_2) - \frac{3\vec{R}[\vec{R} \cdot (m_1\vec{r}_1 + m_2\vec{r}_2)]}{R^2}\right.$$
$$- \frac{3[m_1\vec{r}_1(\vec{R} \cdot \vec{r}_1) + m_2\vec{r}_2(\vec{R} \cdot \vec{r}_2)]}{R^2} - \frac{3}{2}\frac{\vec{R}(m_1 r_1^2 + m_2 r_2^2)}{R^2}$$
$$\left. + \frac{15}{2}\frac{\vec{R}[m_1(\vec{R} \cdot \vec{r}_1)^2 + m_2(\vec{R} \cdot \vec{r}_2)^2]}{R^4}\right\} \qquad (5.83)$$

Knowing

$$m_1\vec{r}_1 + m_2\vec{r}_2 = 0 \qquad (5.84)$$

and

$$\vec{r}_1 = -\gamma\vec{L}, \quad \vec{r}_2 = (1-\gamma)\vec{L}, \quad \vec{L} = L\hat{i}, \quad \vec{R} = R\hat{i}_o \qquad (5.85)$$

we write

$$m_1 r_1^2 + m_2 r_2^2 = M_e L^2 \qquad (5.86)$$

where $M_e = m_1 m_2/(m_1 + m_2)$. Using the preceding relation in Eq. (5.83) results in

$$\vec{F} = -\frac{\mu}{R^3}\left\{ M\vec{R} - \frac{3}{2R}M_e L^2 [2(\hat{i}_o \cdot \hat{i})\hat{i} - 5(\hat{i}_o \cdot \hat{i})^2 \hat{i}_o + \hat{i}_o] \right\} \qquad (5.87)$$

where $M = m_1 + m_2$.

For this problem of three-dimensional motion of the system, the unit vector $\hat{i}$ in the body-fixed rotating frame S_{xyz} is related to the unit vectors in the orbital reference frame $S_{x_o y_o z_o}$ as follows

$$\hat{i} = cos\beta cos\eta\hat{i}_o + sin\beta cos\eta\hat{j}_o - sin\eta\hat{k}_o \qquad (5.88)$$

Applying the above relation, we have the resultant force vector on the system as

$$\vec{F} = -\frac{\mu}{R^3}\left(MR\hat{i}_o - \frac{3}{2R}M_e L^2 \{ -[3cos^2\beta cos^2\eta]\hat{i}_o + [sin2\beta cos^2\eta]\hat{j}_o \right.$$

$$\left. + \hat{i}_o - [cos\beta sin2\eta]\hat{k}_o \} \right) \qquad (5.89)$$

Alternatively, the force $\vec{F}$ can derived from the system potential energy. It is the gradient of potential energy U, given by

$$\vec{F} = -\nabla U \qquad (5.90)$$

where ∇ denotes gradient. The gradient of a function f is defined mathematically as

$$\nabla f(x,y,z) = \frac{\partial f}{\partial x}\hat{i} + \frac{\partial}{\partial y}\hat{j} + \frac{\partial U}{\partial z}\hat{k} \qquad (5.91)$$

where x, y, z are the components along unit vectors $\hat{i}$, $\hat{j}$, and $\hat{k}$, respectively. These unit vectors represent axes of a right-handed orthogonal coordinate frame. Note the variable f is the function of x, y, and z, only.

The gradient of the potential energy U can be written as

$$\nabla U = \frac{\partial U}{\partial x}\hat{i} + \frac{\partial U}{\partial y}\hat{j} + \frac{\partial U}{\partial z}\hat{k} \qquad (5.92)$$

Here $\hat{i}$, $\hat{j}$, and $\hat{k}$ are the unit vectors of the body-fixed coordinate frame $S - xyz$. We write $\vec{R}$ as

$$\vec{R} = R\hat{i}_o \qquad (5.93)$$

Expressing $\hat{i}_o$ in terms of unit vectors in the body-fixed coordinate frame $S - xyz$ using the the transformation matrix

$$\left\{ \begin{array}{c} \hat{i} \\ \hat{j} \\ \hat{k} \end{array} \right\}_{S_{xyz}} = R_{zy}(\beta, \eta) \left\{ \begin{array}{c} \hat{i}_o \\ \hat{j}_o \\ \hat{k}_o \end{array} \right\}_{S_{x_o y_o z_o}} \qquad (5.94)$$

with $R_{zy}(\beta, \eta)$ given by

$$R_{zy}(\beta, \eta) = R_z(\eta)R_z(\beta) = \begin{bmatrix} cos\beta cos\eta & sin\beta cos\eta & -sin\eta \\ -sin\beta & cos\beta & 0 \\ cos\beta sin\eta & sin\beta sin\eta & cos\eta \end{bmatrix} \qquad (5.95)$$

we have

$$\vec{R} = Rcos\beta cos\eta\hat{i} - Rsin\beta\hat{j} + Rcos\beta sin\eta\hat{k} \qquad (5.96)$$

Thus, we write

$$x = Rcos\beta cos\eta, \quad y = -Rsin\beta, \quad z = Rcos\beta sin\eta \qquad (5.97)$$

As we are required to derive the force $\vec{F}$ along unit vectors $\hat{i}_o$, $\hat{j}_o$, and $\hat{k}_o$ in the local-vertical coordinate frame $S - x_o y_o z_o$, we express Eq. (5.92) in terms of unit vectors $\hat{i}_o$, $\hat{j}_o$, and $\hat{k}_o$

$$\nabla U = \left[\frac{\partial U}{\partial x}cos\beta cos\eta - \frac{\partial U}{\partial y}sin\beta + \frac{\partial U}{\partial z}cos\beta sin\eta \right] \hat{i}_o$$

$$+ \left[\frac{\partial U}{\partial x}sin\beta cos\eta + \frac{\partial U}{\partial y}cos\beta + \frac{\partial U}{\partial z}sin\beta sin\eta \right] \hat{j}_o$$

$$+ \left[-\frac{\partial U}{\partial x}sin\eta + \frac{\partial U}{\partial z}cos\eta \right] \hat{k}_o \qquad (5.98)$$

From the potential energy expression,

$$U = f(R, \beta, \eta) \tag{5.99}$$

So, we have to express ∇U with respect to R, β, and η.

$$\frac{\partial U}{\partial R} = \frac{\partial U}{\partial x}\frac{\partial x}{\partial R} + \frac{\partial U}{\partial y}\frac{\partial y}{\partial R} + \frac{\partial U}{\partial z}\frac{\partial z}{\partial R}$$

$$= -R\left[\frac{\partial U}{\partial x}\cos\beta\cos\eta - \frac{\partial U}{\partial y}\sin\beta + \frac{\partial U}{\partial z}\cos\beta\sin\eta\right] \tag{5.100}$$

$$\frac{\partial U}{\partial \beta} = \frac{\partial U}{\partial x}\frac{\partial x}{\partial \beta} + \frac{\partial U}{\partial y}\frac{\partial y}{\partial \beta} + \frac{\partial U}{\partial z}\frac{\partial z}{\partial \beta}$$

$$= -R\left[\frac{\partial U}{\partial x}\sin\beta\cos\eta + \frac{\partial U}{\partial y}\cos\eta + \frac{\partial U}{\partial z}\sin\beta\sin\eta\right] \tag{5.101}$$

$$\frac{\partial U}{\partial \eta} = \frac{\partial U}{\partial x}\frac{\partial x}{\partial \eta} + \frac{\partial U}{\partial y}\frac{\partial y}{\partial \eta} + \frac{\partial U}{\partial z}\frac{\partial z}{\partial \eta}$$

$$= R\cos\beta\left[\frac{\partial U}{\partial x}\sin\eta + \frac{\partial U}{\partial z}\cos\eta\right] \tag{5.102}$$

Thus, we obtain

$$\nabla U = \frac{\partial U}{\partial R}\hat{i}_o - \frac{1}{R}\frac{\partial U}{\partial \beta}\hat{j}_o + \frac{1}{R\cos\beta}\frac{\partial U}{\partial \eta}\hat{k}_o \tag{5.103}$$

Alternatively, we can obtain the same equation for ∇U by first deriving

$$\frac{\partial U}{\partial X} = \frac{\partial U}{\partial R}\frac{\partial R}{\partial X} + \frac{\partial U}{\partial \beta}\frac{\partial \beta}{\partial X} + \frac{\partial U}{\partial \eta}\frac{\partial \eta}{\partial X}, \quad X = x, y, z \tag{5.104}$$

and substitute these derivatives in Eq. (5.98).
Knowing

$$U = -\frac{\mu M}{R} + \frac{\mu}{2R^3}M_e(1 - 3\cos^2\beta\cos^2\eta)L^2$$

the force $\vec{F}$ is obtained as

$$\vec{F} = -\nabla U = -\left[\frac{\mu}{R^2}M - \frac{3}{2}\frac{\mu}{R^4}M_e(1 - 3\cos^2\beta\cos^2\eta)L^2\right]\hat{i}_o$$

$$- \left[\frac{3}{2}\frac{\mu}{R^4}M_eL^2\sin2\beta\cos^2\eta\right]\hat{j}_o - \left[\frac{3}{2}\frac{\mu}{R^4}M_eL^2\sin2\eta\cos\beta\right]\hat{k}_o$$

$$\tag{5.105}$$

Referring to the solution of Problem set 5.1, the external torque due to gravitational force acting on the system is derived as

$$\vec{T} = -\frac{3\mu}{R^3}M_eL^2\left[\sin\eta\cos\eta\cos^2\beta\hat{j} + \sin\beta\cos\beta\cos\eta\hat{k}\right] \tag{5.106}$$

Using the preceding equations for $d\vec{p}/dt$, $d\vec{H}/dt$, $\vec{F}$ and $\vec{T}$ into Eqs. (5.72)-(5.73), the equations of motion of the system are obtained as

$$M(\ddot{R} - R\dot{\theta}^2) = -\frac{\mu M}{R^2} + \frac{3\mu}{2R^4}M_eL^2[1 - 3cos^2\beta cos^2\eta] \qquad (5.107)$$

$$M(R\ddot{\theta} + 2\dot{R}\dot{\theta}) = \frac{3\mu}{2R^4}M_eL^2 sin2\beta cos^2\eta M_eL^2 \qquad (5.108)$$

$$M_eL^2[(\ddot{\theta} + \ddot{\beta})cos\eta - 2(\dot{\theta} + \dot{\beta})\dot{\eta}sin\eta] = -\frac{3\mu}{R^3}M_eL^2 sin\beta cos\beta cos\eta$$
$$\qquad (5.109)$$

$$M_eL^2\left[\ddot{\eta} + (\dot{\theta} + \dot{\beta})^2 sin\eta cos\eta\right] = -\frac{3\mu}{R^3}M_eL^2 sin\eta cos\eta cos^2\beta \qquad (5.110)$$

or

$$\ddot{R} - R\dot{\theta}^2 + \frac{\mu}{R^2} - \frac{3\mu}{2R^2}\frac{M_e}{M}\frac{L^2}{R^2}[1 - 3cos^2\beta cos^2\eta] = 0 \qquad (5.111)$$

$$\ddot{\theta} + 2\frac{\dot{R}}{R}\dot{\theta}) - \frac{3\mu}{2R^3}\frac{M_e}{M}\frac{L^2}{R^2}sin2\beta cos^2\eta = 0 \qquad (5.112)$$

$$\ddot{\beta} - \frac{\dot{R}}{R}\dot{\theta} - 2(\dot{\theta} + \dot{\beta})\dot{\eta}tan\eta + \frac{3}{2}\frac{\mu}{R^3}\left[sin2\beta + \frac{M_e}{M}\frac{L^2}{R^2}sin2\beta cos^2\eta\right] = 0 \qquad (5.113)$$

$$\ddot{\eta} + \frac{1}{2}(\dot{\theta} + \dot{\beta})^2 sin2\eta + \frac{3}{2}\frac{\mu}{R^3}sin2\eta cos^2\beta = 0 \qquad (5.114)$$

Lagrange Method

The equations of motion of the system with generalized coordinates $q_1 = R$, $q_2 = \theta$, $q_3 = \beta$, and $q_4 = \eta$ are given by

$$\frac{d}{dt}\left(\frac{\partial T}{\partial \dot{q}_k}\right) - \frac{\partial T}{\partial q_k} + \frac{\partial U}{\partial q_k} = 0, \quad k = 1, 2, 3, 4 \qquad (5.115)$$

The potential and kinetic energies of the system are

$$U = -\frac{\mu M}{R} + \frac{\mu}{2R^3}M_e(1 - 3cos^2\beta cos^2\eta)L^2 \qquad (5.116)$$

$$T = \frac{1}{2}M(\dot{R}^2 + \dot{\theta}^2R^2) + \frac{1}{2}M_e[(\dot{\theta} + \dot{\beta})^2 cos^2\eta + \dot{\eta}^2]L^2 \qquad (5.117)$$

Using the preceding equations, the equations of motion of the system are derived as follows:

<u>R-equation</u>

$$\frac{d}{dt}\left(\frac{\partial T}{\partial \dot{R}}\right) - \frac{\partial T}{\partial R} + \frac{\partial U}{\partial R} = 0 \qquad (5.118)$$

Here

$$\frac{\partial T}{\partial \dot{R}} = M\dot{R}, \Rightarrow \frac{d}{dt}\left(\frac{\partial T}{\partial \dot{R}}\right) = M\ddot{R} \qquad (5.119)$$

$$\frac{\partial T}{\partial R} = M\dot{\theta}^2 R \qquad (5.120)$$

$$\frac{\partial U}{\partial R} = \frac{\mu M}{R^2} - \frac{3\mu}{2R^4}M_e(1 - 3cos^2\beta cos^2\eta)L^2 \qquad (5.121)$$

Thus, the R-equation of motion is

$$M\ddot{R} - M\dot{\theta}^2 R + \frac{\mu M}{R^2} - \frac{3\mu}{2R^4}M_e(1 - 3cos^2\beta cos^2\eta)L^2 = 0 \qquad (5.122)$$

θ-equation

$$\frac{d}{dt}\left(\frac{\partial T}{\partial \dot{\theta}}\right) - \frac{\partial T}{\partial \theta} + \frac{\partial U}{\partial \theta} = 0 \qquad (5.123)$$

Here

$$\frac{\partial T}{\partial \dot{\theta}} = M\dot{\theta}R^2 + M_e L^2(\dot{\theta} + \dot{\beta})cos^2\eta \qquad (5.124)$$

$$\Rightarrow \frac{d}{dt}\left(\frac{\partial T}{\partial \dot{\theta}}\right) = M\ddot{\theta}R^2 + 2M\dot{\theta}R\dot{R} + M_e L^2[(\ddot{\theta} + \ddot{\beta})cos^2\eta - (\dot{\theta} + \dot{\beta})\dot{\eta}sin2\eta] \qquad (5.125)$$

$$\frac{\partial T}{\partial \theta} = 0, \quad \frac{\partial U}{\partial \theta} = 0 \qquad (5.126)$$

Thus, the θ-equation of motion is

$$M\ddot{\theta}R^2 + 2M\dot{\theta}R\dot{R} + M_e L^2[(\ddot{\theta} + \ddot{\beta})cos^2\eta - (\dot{\theta} + \dot{\beta})\dot{\eta}sin2\eta] = 0 \qquad (5.127)$$

β-equation

$$\frac{d}{dt}\left(\frac{\partial T}{\partial \dot{\beta}}\right) - \frac{\partial T}{\partial \beta} + \frac{\partial U}{\partial \beta} = 0 \qquad (5.128)$$

Here

$$\frac{\partial T}{\partial \dot{\beta}} = M_e L^2(\dot{\theta} + \dot{\beta})cos^2\eta \qquad (5.129)$$

$$\Rightarrow \frac{d}{dt}\left(\frac{\partial T}{\partial \dot{\beta}}\right) = M_e L^2[(\ddot{\theta} + \ddot{\beta})cos^2\eta - (\dot{\theta} + \dot{\beta})\dot{\eta}sin2\eta] \qquad (5.130)$$

$$\frac{\partial T}{\partial \beta} = 0 \qquad (5.131)$$

$$\frac{\partial U}{\partial \beta} = \frac{3\mu}{R^3}M_e L^2 cos\beta sin\beta cos^2\eta \qquad (5.132)$$

Thus, the β-equation of motion is

$$M_e L^2[(\ddot{\theta}+\ddot{\beta})cos^2\eta - (\dot{\theta}+\dot{\beta})\dot{\eta}sin2\eta] + \frac{3\mu}{R^3}M_e L^2 cos\beta sin\beta cos^2\eta = 0$$

(5.133)

η-equation

$$\frac{d}{dt}\left(\frac{\partial T}{\partial \dot{\eta}}\right) - \frac{\partial T}{\partial \eta} + \frac{\partial U}{\partial \eta} = 0$$

(5.134)

Here

$$\frac{\partial T}{\partial \dot{\eta}} = M_e \dot{\eta}L^2$$

(5.135)

$$\Rightarrow \frac{d}{dt}\left(\frac{\partial T}{\partial \dot{\eta}}\right) = M_e \ddot{\eta}L^2$$

(5.136)

$$\frac{\partial T}{\partial \eta} = -M_e L^2(\dot{\theta}+\dot{\beta})^2 cos\eta sin\eta$$

(5.137)

$$\frac{\partial U}{\partial \eta} = \frac{3\mu}{R^3}M_e L^2 cos\eta sin\eta cos^2\beta$$

(5.138)

Thus, the η-equation of motion is

$$M_e L^2\ddot{\eta} + M_e L^2(\dot{\theta}+\dot{\beta})^2 cos\eta sin\eta + \frac{3\mu}{R^3}M_e L^2 cos\eta sin\eta cos^2\beta = 0$$

(5.139)

Thus, the equations of motion of the system are

$$M\ddot{R} - M\dot{\theta}^2 R + \frac{\mu M}{R^2} - \frac{3\mu}{2R^4}M_e(1 - 3cos^2\beta cos^2\eta)L^2 = 0$$

(5.140)

$$M\ddot{\theta}R^2 + 2M\dot{\theta}R\dot{R} + M_e L^2[(\ddot{\theta}+\ddot{\beta})cos^2\eta - (\dot{\theta}+\dot{\beta})\dot{\eta}sin2\eta] = 0$$ (5.141)

$$M_e L^2[(\ddot{\theta}+\ddot{\beta})cos^2\eta - (\dot{\theta}+\dot{\beta})\dot{\eta}sin2\eta] + \frac{3\mu}{R^3}M_e L^2 cos\beta sin\beta cos^2\eta = 0$$

(5.142)

$$M_e L^2\ddot{\eta} + M_e L^2(\dot{\theta}+\dot{\beta})^2 cos\eta sin\eta + \frac{3\mu}{R^3}M_e L^2 cos\eta sin\eta cos^2\beta = 0$$

(5.143)

or

$$\ddot{R} - R\dot{\theta}^2 + \frac{\mu}{R^2} - \frac{3\mu}{2R^2}\frac{M_e}{M}\frac{L^2}{R^2}[1 - 3cos^2\beta cos^2\eta] = 0 \qquad (5.144)$$

$$\ddot{\theta} + 2\frac{\dot{R}}{R}\dot{\theta}) - \frac{3\mu}{2R^3}\frac{M_e}{M}\frac{L^2}{R^2}sin2\beta cos^2\eta = 0 \qquad (5.145)$$

$$\ddot{\beta} - \frac{\dot{R}}{R}\dot{\theta} - 2(\dot{\theta} + \dot{\beta})\dot{\eta}tan\eta + \frac{3}{2}\frac{\mu}{R^3}\left[sin2\beta + \frac{M_e}{M}\frac{L^2}{R^2}sin2\beta cos^2\eta\right] = 0$$
$$(5.146)$$

$$\ddot{\eta} + \frac{1}{2}(\dot{\theta} + \dot{\beta})^2 sin2\eta + \frac{3}{2}\frac{\mu}{R^3}sin2\eta cos^2\beta = 0 \qquad (5.147)$$

Chapter 6

Mathematical and Numerical Simulation

Problem Set 6

6.1 The linear equations of motion of the system are obtained as

$$\delta\beta'' + 3\delta\beta = 0 \qquad (6.1)$$
$$\delta\eta'' + 4\delta\eta = 0 \qquad (6.2)$$

6.2 The linear equations of motion of the system are obtained as

$$(M + m)\ddot{x} - ml\ddot{\theta} = -c\dot{x} + u \qquad (6.3)$$
$$ml\ddot{x} - (J + ml^2)\ddot{\theta} = mgl\theta \qquad (6.4)$$

or in state space form:

$$\dot{X} = AX + Bu \qquad (6.5)$$

where

$$X = \begin{bmatrix} x \\ \dot{x} \\ \theta \\ \dot{\theta} \end{bmatrix}, \quad A = \begin{bmatrix} 0 & 0 & 1 & 0 \\ 0 & \dfrac{gm^2l^2}{J(M+m)+Mml^2} & \dfrac{-(J+ml^2)b}{J(M+m)+Mml^2} & 0 \\ 0 & 0 & 0 & 1 \\ 0 & \dfrac{mgl(M+m)}{J(M+m)+Mml^2} & \dfrac{-mlb}{J(M+m)+Mml^2} & 0 \end{bmatrix},$$

$$B = \begin{bmatrix} 0 \\ \dfrac{J+ml^2}{J(M+m)+Mml^2} \\ 0 \\ \dfrac{ml}{J(M+m)+Mml^2} \end{bmatrix}$$

6.3 The characteristic equation of the given system

$$\ddot{\alpha} + 5\dot{\alpha} + 25\alpha = 0 \tag{6.6}$$

is

$$\lambda^2 + 5\lambda + 25 = 0 \tag{6.7}$$

Here λ corresponds to the characteristic root or eigenvalue of the system.

The preceding equation is compared with the standard second-order system:

$$\ddot{x} + 2\zeta\omega_n\dot{x} + \omega_n^2 x = 0 \tag{6.8}$$

and its characteristic equation given as

$$\lambda^2 + 2\zeta\omega_n\lambda + \omega_n^2 = 0 \tag{6.9}$$

where ω_n is the undamped natural frequency and ζ is the damping ratio.

Comparing Eqs. (6.9) and (6.7), we can write

$$2\zeta\omega_n = 5 \qquad (6.10)$$
$$\omega_n^2 = 25 \qquad (6.11)$$

Solving Eqs. (6.11), we get

$$\omega_n = 5 \text{ rad/s} \qquad (6.12)$$
$$\zeta = \frac{5}{2\omega_n} = \frac{5}{10} = 0.5 \qquad (6.13)$$

The roots of the characteristic Eq. (6.9) are

$$\lambda_{1,2} = -\zeta\omega_n \pm j\omega_n\sqrt{1-\zeta^2}$$
$$= -\zeta\omega_n \pm j\omega_d \qquad (6.14)$$

Here $\omega_d = \omega_n\sqrt{1-\zeta^2}$ is called damped natural frequency of the system. So, ω_d for the given system of Eq. (6.6) is

$$\omega_d = \omega_n\sqrt{1-\zeta^2} = 5 \times \sqrt{1-0.5^2} = 4.33 \text{ rad/s} \qquad (6.15)$$

Thus, we have the following answers:

(a) $\omega_n = 5$ rad/s

(b) $\zeta = 0.5$

(c) $\omega_d = 4.33$ rad/s

6.4 Given the second-order differential equation

$$\ddot{\theta} + 2\dot{\theta} + 5\theta = -\delta$$

we can write as first-order differential equations

$$\dot{x}_1 = x_2$$
$$\dot{x}_2 = -2x_2 - 5x_1 - \delta$$

where $x_1 = \theta$, and $x_2 = \dot{\theta}$.

In state space form:

$$\left\{ \begin{array}{c} \dot{x}_1 \\ \dot{x}_2 \end{array} \right\} = \begin{bmatrix} 0 & 1 \\ -5 & -2 \end{bmatrix} \left\{ \begin{array}{c} x_1 \\ x_2 \end{array} \right\} + \begin{bmatrix} 0 \\ -1 \end{bmatrix} \delta$$

where

$$A = \begin{bmatrix} 0 & 1 \\ -5 & -2 \end{bmatrix}$$

$$B = \begin{bmatrix} 0 \\ -1 \end{bmatrix} \delta$$

The eigenvalues of the system can be determined by solving the equation

$$|A - \lambda I| = 0$$

where I is the identity matrix. Substituting the A matrix into the preceding equation yields

$$\left| \begin{bmatrix} 0 & 1 \\ -5 & -2 \end{bmatrix} - \lambda \begin{bmatrix} 1 & 0 \\ 0 & 1 \end{bmatrix} \right| = \left| \begin{bmatrix} 0 & 1 \\ -5 & -2 \end{bmatrix} - \begin{bmatrix} \lambda & 0 \\ 0 & \lambda \end{bmatrix} \right| = 0$$

or

$$\begin{vmatrix} -\lambda & 1 \\ -5 & -2 - \lambda \end{vmatrix} = 0$$

Expanding the determinant yields the characteristic equation

$$\lambda(\lambda + 2) + 5 = 0$$

or

$$\lambda^2 + 2\lambda + 5 = 0$$

The characteristic equation can be solved for the eigenvalues for the system.

The eigenvalues for this particular characteristic equation are

$$\lambda_{1,2} = -1 \pm 2j$$

The eigenvalues are complex and the real part of the root is negative. This means that the system is dynamically stable. If the system were given an initial disturbance, the motion would decay sinusoidally and the frequency of the oscillation would be governed by the imaginary part of the complex eigenvalues.

6.7 Solving the differential equations for the highest order derivative yield

$$\dot{x}_1 = -0.5x_1 + 10x_2 - \delta$$
$$\dot{x}_2 = -x_1 + x_2 + 2\delta$$

or in matrix form

$$\left\{ \begin{array}{c} \dot{x}_1 \\ \dot{x}_2 \end{array} \right\} = \begin{bmatrix} -0.5 & 10 \\ -1.0 & 1.0 \end{bmatrix} \left\{ \begin{array}{c} x_1 \\ x_2 \end{array} \right\} + \begin{bmatrix} -1 \\ 2 \end{bmatrix} \delta$$

which is the state space formulation

$$\dot{\mathbf{x}} = \mathbf{A}\mathbf{x} + \mathbf{B}\mathbf{u}$$

where

$$A = \begin{bmatrix} -0.5 & 10 \\ -1.0 & 1.0 \end{bmatrix}$$

$$B = \begin{bmatrix} -1 \\ 2 \end{bmatrix} \delta$$

The eigenvalues of the system can be determined by solving the equation

$$|\lambda I - A| = 0$$

where I is the identity matrix. Substituting the A matrix into the preceding equation yields

$$\left| \lambda \begin{bmatrix} 1 & 0 \\ 0 & 1 \end{bmatrix} - \begin{bmatrix} -0.5 & 10 \\ -1.0 & 1.0 \end{bmatrix} \right| = 0$$

$$\left| \begin{bmatrix} \lambda & 0 \\ 0 & \lambda \end{bmatrix} - \begin{bmatrix} -0.5 & 10 \\ -1.0 & 1.0 \end{bmatrix} \right| = 0$$

$$\begin{vmatrix} \lambda + 0.5 & -10 \\ 1.0 & \lambda - 1.0 \end{vmatrix} = 0$$

Expanding the determinant yields the characteristic equation

$$(\lambda + 0.5)(\lambda - 1.0) + 10 = 0$$

or

$$\lambda^2 - 0.5\lambda + 9.5 = 0$$

The characteristic equation can be solved for the eigenvalues for the system.

The eigenvalues for this particular characteristic equation are

$$\lambda_{1,2} = 0.25 \pm 3.07j$$

The eigenvalues are complex and the real part of the root is positive. This means that the system is dynamically unstable. If the system were given an initial disturbance, the motion would grow sinusoidally and the frequency of the oscillation would be governed by the imaginary part of the complex eigenvalues.

6.8 Given the differential equation

$$\dddot{x} + \ddot{x} - 4\dot{x} + 6x = r$$

we can write the first-order differential equations as

$$\dot{x}_1 = x_2$$
$$\dot{x}_2 = x_3$$
$$\dot{x}_3 = r - 6x_1 + 4x_2 - x_3$$

where $x_1 = x$, $x_2 = \dot{x}$, and $x_3 = \ddot{x}$.

In state space form:

$$\left\{ \begin{array}{c} \dot{x}_1 \\ \dot{x}_2 \\ \dot{x}_3 \end{array} \right\} = \begin{bmatrix} 0 & 1 & 0 \\ 0 & 0 & 1 \\ -6 & 4 & -1 \end{bmatrix} \left\{ \begin{array}{c} x_1 \\ x_2 \\ x_3 \end{array} \right\} + \begin{bmatrix} 0 \\ 0 \\ 1 \end{bmatrix} r$$

where

$$A = \begin{bmatrix} 0 & 1 & 0 \\ 0 & 0 & 1 \\ -6 & 4 & -1 \end{bmatrix}$$

$$B = \begin{bmatrix} 0 \\ 0 \\ 1 \end{bmatrix} r$$

The eigenvalues of the system can be determined by solving the equation

$$|A - \lambda I| = 0$$

where I is the identity matrix. Substituting the A matrix into the preceding equation yields

$$\left| \begin{bmatrix} 0 & 1 & 0 \\ 0 & 0 & 1 \\ -6 & 4 & -1 \end{bmatrix} - \lambda \begin{bmatrix} 1 & 0 & 0 \\ 0 & 1 & 0 \\ 0 & 0 & 1 \end{bmatrix} \right| = \left| \begin{bmatrix} 0 & 1 & 0 \\ 0 & 0 & 1 \\ -6 & 4 & -1 \end{bmatrix} - \begin{bmatrix} \lambda & 0 & 0 \\ 0 & \lambda & 0 \\ 0 & 0 & \lambda \end{bmatrix} \right| = 0$$

or

$$\begin{vmatrix} -\lambda & 1 & 0 \\ 0 & -\lambda & 1 \\ -6 & 4 & -1-\lambda \end{vmatrix} = 0$$

Expanding the determinant yields the characteristic equation

$$-\lambda[-\lambda(-1-\lambda)-4]-1[0+6]=0$$

or

$$\lambda^3 + \lambda^2 - 4\lambda + 6 = 0$$

The characteristic equation can be solved for the eigenvalues for the system.

The eigenvalues for this particular characteristic equation are

$$\lambda_1 = -3$$
$$\lambda_{2,3} = 1 \pm j$$

As the real part of one of the eigenvalues is positive, the system is dynamically unstable. If the system were given an initial disturbance, the motion would grow sinusoidally and the frequency of the oscillation would be governed by the imaginary part of the complex eigenvalues.

6.9 For the given linear equations of motion of a system

$$\dot{\mathbf{x}} = \mathbf{A}\mathbf{x} + \mathbf{B}\mathbf{u} \tag{6.16}$$

the general solution is given by

$$\mathbf{x} = \mathbf{x_0} e^{\lambda t} \tag{6.17}$$

where $\mathbf{x_0}$=initial state $\mathbf{x}$ at $t=0$. λ=eigenvalue or root of the characteristic equation of the system and it comprises of a real part η and an imaginary part ω as

$$\lambda = \eta + j\omega \qquad (6.18)$$

The real part η states about the magnitude of $\mathbf{x}$ whereas the imaginary part ω says about an oscillatory motion of $\mathbf{x}$. As we are interested in magnitude of $\mathbf{x}$, we take only the real part, *i.e.*,

$$\lambda = \eta \qquad (6.19)$$

Putting λ in Eq. (6.17), and solving for time t by taking logarithm with base e both sides

$$ln\frac{\mathbf{x}}{\mathbf{x_0}} = \eta t$$

or

$$t = \frac{1}{\eta}ln\frac{\mathbf{x}}{\mathbf{x_0}} \qquad (6.20)$$

As per Eqs. (6.17) and (6.19), the magnitude of $\mathbf{x}$ will increase if η is positive. Thus, substituting $\mathbf{x} = 2\mathbf{x_0}$ in Eq. (6.20) for doubling of the amplitude with positive η, the corresponding time t_2 can be written as

$$t_2 = \frac{1}{\eta}ln\frac{2\mathbf{x_0}}{\mathbf{x_0}} = \frac{1}{\eta}ln2 = \frac{0.693}{\eta} \qquad (6.21)$$

Similarly, the magnitude of $\mathbf{x}$ will decrease if η is negative as per Eqs. (6.17) and (6.19), Thus, for having of the amplitude we substitute $\mathbf{x} = \mathbf{x_0}/2$ in Eq. (6.20) and the corresponding time $t_{1/2}$ is

$$t_{1/2} = \frac{1}{\eta}ln\frac{\mathbf{x_0}/2}{\mathbf{x_0}} = -\frac{1}{\eta}ln2 = -\frac{0.693}{\eta} \qquad (6.22)$$

Here η will have negative value as explained earlier.

Thus, from Eqs. (6.21) and (6.22) we conclude that the time for doubling or halving of the amplitude is

$$t_2 \quad or \quad t_{1/2} = \frac{0.693}{|\eta|} \qquad (6.23)$$

The corresponding number of cycles N_2 and $N_{1/2}$ can be given by

$$N(cycles)_2 = \frac{t_2}{T}$$
$$N(cycles)_{1/2} = \frac{t_{1/2}}{T} \tag{6.24}$$

where

$$T = \text{time period} = \frac{2\pi}{\omega} \tag{6.25}$$

Using Eqs. (6.23) and (6.25), Eqs. (6.24) can be rewritten as

$$N(cycles)_2 \quad or \quad N(cycles)_{1/2} = \frac{0.693|\omega|}{|\eta|2\pi} = 0.110\frac{|\omega|}{|\eta|} \tag{6.26}$$

6.10 For a given characteristic equation of the system, the eigenvalues obtained are

$$\lambda_{1,2} = -0.0171 \pm j0.213$$
$$\lambda_{3,4} = -2.5 \pm j2.59$$

the system is dynamically stable as all real parts of the roots are negative.

For the given system, T (period), $t_{1/2}$ (time to half amplitude), and $N_{1/2}$ (number of cycles to half amplitude) corresponding to $\lambda_{1,2}$ and $\lambda_{3,4}$ are tabulated below.

$\lambda_{1,2}$	$\lambda_{3,4}$
$\eta = -\zeta\omega_n = -0.0171$	$\eta = -2.5$
$\omega = 0.213 \ rad/s$	$\omega = 2.59 \ rad/s$
$T = \frac{2\pi}{\omega} = \frac{2\pi}{0.213}$	$T = \frac{2\pi}{\omega} = \frac{2\pi}{2.59}$
$T = 29.5 \ s$	$T = 2.42 \ s$
$t_{1/2} = \frac{0.693}{|\eta|} = \frac{0.693}{0.0171}$	$t_{1/2} = \frac{0.693}{|\eta|} = \frac{0.693}{2.5}$
$t_{1/2} = 40.3 \ s$	$t_{1/2} = 0.28 \ s$
$N_{1/2} = \frac{t_{1/2}}{T} = \frac{0.11\omega}{|\eta|}$	$N_{1/2} = \frac{t_{1/2}}{T} = \frac{0.11\omega}{|\eta|}$
$N_{1/2} = \frac{0.11 \times 0.213}{0.0171} = 1.37 \ cycles$	$N_{1/2} = \frac{0.11 \times 2.59}{2.5} = 0.11 \ cycles$

6.13 We consider the characteristic equation in the form

$$\lambda^n + a_{n-1}\lambda^{n-1} + \cdots + a_1\lambda + a_0 = 0 \qquad (6.27)$$

where a_j, $j = 0, 1, \ldots, n - 1$ are constant coefficients.

For $n=2$, we write the characteristic equation

$$\lambda^2 + a_1\lambda + a_0 = 0 \qquad (6.28)$$

and its solution is

$$\lambda_{1,2} = -\frac{a_1}{2} \pm \frac{\sqrt{a_1^2 - 4a_0}}{2} \qquad (6.29)$$

The sum of the roots can be written as

$$\lambda_1 + \lambda_2 = -a_1$$

or

$$a_1 = -(\lambda_1 + \lambda_2) \qquad (6.30)$$

Thus, the coefficient of λ_1 is the negative of the sum of all the real parts as well as imaginary parts of the roots.

For $n=3$, the characteristic equation is written as

$$\lambda^3 + a_2\lambda^2 + a_1\lambda + a_0 = 0 \qquad (6.31)$$

Let the roots be

$$\lambda = \eta + j\omega \qquad (6.32)$$

where η=vector of real parts of the roots; ω=vector of imaginary parts of the roots.

As $n=3$, there would be 3 roots and thus the vector η can be considered as

$$\eta = \left\{ \begin{array}{c} \eta_1 \\ \eta_2 \\ \eta_3 \end{array} \right\} \tag{6.33}$$

Substituting η in Eq. (6.31), we can write in matrix form

$$\begin{bmatrix} \eta_1^3 & \eta_1^2 & \eta_1 & 1 \\ \eta_2^3 & \eta_2^2 & \eta_2 & 1 \\ \eta_3^3 & \eta_3^2 & \eta_3 & 1 \end{bmatrix} \left\{ \begin{array}{c} 1 \\ a_2 \\ a_1 \\ a_0 \end{array} \right\} = 0 \tag{6.34}$$

Applying (Row 2 - Row 1) and (Row 3 - Row 1), we get

$$\begin{bmatrix} \eta_1^3 & \eta_1^2 & \eta_1 & 1 \\ \eta_2^3 - \eta_1^3 & \eta_2^2 - \eta_1^2 & \eta_2 - \eta_1 & 0 \\ \eta_3^3 - \eta_1^3 & \eta_3^2 - \eta_1^2 & \eta_3 - \eta_1 & 0 \end{bmatrix} \left\{ \begin{array}{c} 1 \\ a_2 \\ a_1 \\ a_0 \end{array} \right\} = 0 \tag{6.35}$$

Using the relation $a^3 - b^3 = (a - b)\{(a - b)^2 + 3ab\}$, The preceding equation simplifies to

$$(\eta_2 - \eta_1)(\eta_3 - \eta_1) \begin{bmatrix} \eta_1^3 & \eta_1^2 & \eta_1 & 1 \\ \begin{array}{c}(\eta_2 - \eta_1)^2 \\ +3\eta_1\eta_2\end{array} & \eta_1 + \eta_2 & 1 & 0 \\ \begin{array}{c}(\eta_3 - \eta_1)^2 \\ +3\eta_1\eta_3\end{array} & \eta_1 + \eta_3 & 1 & 0 \end{bmatrix} \left\{ \begin{array}{c} 1 \\ a_2 \\ a_1 \\ a_0 \end{array} \right\} = 0 \tag{6.36}$$

Applying (Row 2 - Row 3), we get

$$(\eta_2 - \eta_1)(\eta_3 - \eta_1)(\eta_2 - \eta_3) \begin{bmatrix} \eta_1^3 & \eta_1^2 & \eta_1 & 1 \\ \eta_1 + \eta_2 + \eta_2 & 1 & 1 & 0 \\ (\eta_3 - \eta_1)^2 & \eta_1 + \eta_3 & 1 & 0 \\ +3\eta_1\eta_3 & & & \end{bmatrix} \begin{Bmatrix} 1 \\ a_2 \\ a_1 \\ a_0 \end{Bmatrix} = 0$$

$$(6.37)$$

As $\eta_1 \neq \eta_2 \neq \eta_3$, we can write

$$\begin{bmatrix} \eta_1^3 & \eta_1^2 & \eta_1 & 1 \\ \eta_1 + \eta_2 + \eta_3 & 1 & 1 & 0 \\ (\eta_3 - \eta_1)^2 & \eta_1 + \eta_3 & 1 & 0 \\ +3\eta_1\eta_3 & & & \end{bmatrix} \begin{Bmatrix} 1 \\ a_2 \\ a_1 \\ a_0 \end{Bmatrix} = 0 \qquad (6.38)$$

Thus, from Row 2 we can write

$$\eta_1 + \eta_2 + \eta_3 + a_2 = 0$$

or

$$a_2 = -(\eta_1 + \eta_2 + \eta_3) \qquad (6.39)$$

Thus, a_2 the coefficient of λ_2 is the negative of the sum of the real parts of all the roots.

6.14 The equations of motion of the given system can be written as

$$\ddot{\alpha} = \frac{-a_{12}a_{21}\alpha + a_{11}a_{22}\dot{\beta}_1}{a_{21} - a11}$$

$$\ddot{\beta}_1 = \frac{a_{12}\alpha - a_{22}\dot{\beta}_1}{a_{21} - a11} \qquad (6.40)$$

We can rewrite the system equations of motion in a general form as

$$\dot{\mathbf{x}} = \mathbf{A}\mathbf{x} \qquad (6.41)$$

where

$$\mathbf{x} = \text{state vector} = [\alpha, \ \dot{\alpha}, \ \beta_1, \ \dot{\beta}_1]^T$$

$$A = \frac{1}{a_{21} - a_{11}} \begin{bmatrix} 0 & 1 & 0 & 0 \\ -a_{12}a_{21} & 0 & 0 & a_{11}a_{22} \\ 0 & 0 & 0 & 1 \\ a_{12} & 0 & 0 & -a_{22} \end{bmatrix}$$

As there are four state variables, the order of the characteristic equation is four. However, all elements in the third column of A (corresponding to β_1) are zero. It implies that the state variable β_1 is not involved in the system equations of motion and the corresponding eigenvalue would be zero. So, the system state variables are reduced to $\mathbf{x} = [\alpha, \ \dot{\alpha}, \ \dot{\beta}_1]^T$ and thus, the order of the characteristic equation is three.

6.15 We can write the system equations of motion in a general form as

$$\dot{\mathbf{x}} = \mathbf{A}\mathbf{x} \qquad\qquad (6.42)$$

where

$$\mathbf{x} = \text{state vector} = [\phi, \ \dot{\phi}, \ \gamma, \ \dot{\gamma}, \ \beta_2, \ \dot{\beta}_2, \ \beta_3, \ \dot{\beta}_3]^T$$

As there are eight state variables, the order of the characteristic equation is eight. However, all elements in the fifth and seventh columns of A (corresponding to β_2 and be_3) would be zero. It implies that the state variables β_2 and β_3 are not involved in the system equations of motion and corresponding two eigenvalues would be zero. So, the system state variables are reduced to $\mathbf{x} = [\phi, \ \dot{\phi}, \ \gamma, \ \dot{\gamma}, \ \dot{\beta}_2, \ \dot{\beta}_3]^T$ and thus, the order of the characteristic equation is six.

6.17 Define the state variables as

$$x_1 = r, \ \ x_2 = \dot{r}, \ \ x_3 = \theta, \ \ x_4 = \dot{\theta}$$

then

$$\dot{x}_1 = x_2, \ \ \ \dot{x}_3 = x_4$$

Substituting these in the equations of motion, we can write the first-order equations of motion of the spacecraft as

$$\dot{x}_1 = x_2$$
$$\dot{x}_3 = x_4$$
$$\dot{x}_2 = x_1 x_4^2 - \frac{\mu}{x_1^2} + a_r$$
$$\dot{x}_4 = -2\frac{x_2 x_4}{x_1} + \frac{a_\theta}{x_1}$$

6.18 The state vectors of the system are expressed in terms of x_i, $i = 1, 2, \ldots, 6$: $X = [R, \dot{R}, \theta, \dot{\theta}, \beta, \dot{\beta}] = [x_1, x_2, x_3, x_4, x_5, x_6]$. The Maple program is shown in Fig. 6.1. Note Eq_1, Eq_2, and Eq_3 represent the equations of motion for R, θ, and β degrees of freedom, respectively.

6.19 The state vectors of the system are expressed in terms of x_i, $i = 1, 2, \ldots, 6$: $X = [R, \dot{R}, \theta, \dot{\theta}, \beta, \dot{\beta}, u, \dot{u}] = [x_1, x_2, x_3, x_4, x_5, x_6, x_7, x_8]$. The Maple program is shown in Figs. 6.2-6.5. Note Eq_1, Eq_2, Eq_3, and Eq_4 represent the equations of motion for R, θ, β, and u degrees of freedom, respectively. Nomenclature for mass moment of inertia I_j, $j = x, y, z$ is replaced by J_j, $j = x, y, z$ since I is a reserved notation in Maple.

> $T := \frac{1}{2}M\left(x_2^2 + x_4^2\ x_1^2\right) + \frac{1}{2}\ M_e\ (x_4 + x_6)^2 l^2$

$$T := \frac{1}{2}\ M\left(x_2^2 + x_4^2 x_1^2\right) + \frac{1}{2}\ M_e\ (x_4 + x_6)^2\ l^2 \tag{1}$$

> $U := \frac{-mu\ M}{x_1} + \frac{\mu}{2\ x_1^3}\ M_e\ \left(1 - 3\ cos(x_5)^2\right) l^2$

$$U := -\frac{\mu M}{x_1} + \frac{1}{2}\ \frac{\mu M_e\left(1 - 3\ cos(x_5)^2\right)\ l^2}{x_1^3} \tag{2}$$

> $Lagrange := T - U$

$$Lagrange := \frac{1}{2}\ M\left(x_2^2 + x_4^2 x_1^2\right) + \frac{1}{2}\ M_e\ (x_4 + x_6)^2\ l^2 + \frac{\mu M}{x_1} - \frac{1}{2}\ \frac{\mu M_e\left(1 - 3\ cos(x_5)^2\right)\ l^2}{x_1^3} \tag{3}$$

> $g := diff\left(Lagrange, x_2\right)$

$$g := Mx_2 \tag{4}$$

> $g_t := subs\left(\ \{x_1 = x_1(t), x_2 = x_2(t), x_3 = x_3(t), x_4 = x_4(t), x_5 = x_5(t), x_6 = x_6(t)\}, g\right)$

$$g_t := Mx_2(t) \tag{5}$$

> $Eq_1 := diff\left(g_t, t\right) - diff\left(Lagrange, x_1\right) = 0$

$$Eq_1 := M\left(\frac{d}{dt}\ x_2(t)\right) - Mx_4^2 x_1 + \frac{\mu M}{x_1^2} - \frac{3}{2}\ \frac{\mu M_e\left(1 - 3\ cos(x_5)^2\right)\ l^2}{x_1^4} = 0 \tag{6}$$

> $g := diff\left(Lagrange, x_4\right)$

$$g := Mx_4 x_1^2 + M_e\left(x_4 + x_6\right) l^2 \tag{7}$$

> $g_t := subs\left(\ \{x_1 = x_1(t), x_2 = x_2(t), x_3 = x_3(t), x_4 = x_4(t), x_5 = x_5(t), x_6 = x_6(t)\}, g\right)$

$$g_t := Mx_4(t)\, x_1(t)^2 + M_e\left(x_4(t) + x_6(t)\right) l^2 \tag{8}$$

> $Eq_2 := diff\left(g_t, t\right) - diff\left(Lagrange, x_3\right) = 0$

$$Eq_2 := M\left(\frac{d}{dt}\ x_4(t)\right) x_1(t)^2 + 2\, Mx_4(t)\, x_1(t)\left(\frac{d}{dt}\ x_1(t)\right) + M_e\left(\frac{d}{dt}\ x_4(t) + \frac{d}{dt}\ x_6(t)\right) l^2 = 0 \tag{9}$$

> $g := diff\left(Lagrange, x_6\right)$

$$g := M_e\left(x_4 + x_6\right) l^2 \tag{10}$$

> $g_t := subs\left(\ \{x_1 = x_1(t), x_2 = x_2(t), x_3 = x_3(t), x_4 = x_4(t), x_5 = x_5(t), x_6 = x_6(t)\}, g\right)$

$$g_t := M_e\left(x_4(t) + x_6(t)\right) l^2 \tag{11}$$

> $Eq_3 := diff\left(g_t, t\right) - diff\left(Lagrange, x_5\right) = 0$

$$Eq_3 := M_e\left(\frac{d}{dt}\ x_4(t) + \frac{d}{dt}\ x_6(t)\right) l^2 + \frac{3\ \mu M_e cos(x_5)\ sin(x_5)\ l^2}{x_1^3} = 0 \tag{12}$$

Figure 6.1: Maple program for Problem 6.18.

> T := $\frac{1}{2}$ M $\left(x_2^2 + x_4^2\, x_1^2\right) + \frac{1}{2}$ m_2 $\left(w^2\, a^2 + x_{10}^2 + w_l\,^2 l^2\right.$

$\qquad + 2\ a\ w\left(-x_{10}\ sin\left(x_5 - x_7\right) + w_l\ l\ cos\left(x_5 - x_7\right)\right)\Big) + \frac{1}{2}\ J_z\, w^2$

$T := \frac{1}{2}\, M\left(x_2^2 + x_4^2 x_1^2\right) + \frac{1}{2}\, m_2\left(w^2 a^2 + x_{10}^2 + w_l^2\, l^2 + 2\, a\, w\left(-x_{10}\sin\left(x_5 - x_7\right)\right.\right.$ (1)

$\qquad \left.\left. + w_l\, l\cos\left(x_5 - x_7\right)\right)\right) + \frac{1}{2}\, J_z\, w^2$

> U := $\frac{-mu\ M}{x_1} + \frac{\mu}{2\, x_1^{\,3}}$ m_2 $\left(a^2 + l^2\right.$

$\qquad + 2\ a\ l\ cos\left(x_5 - x_7\right) - 3\ \left(a\ cos\left(x_5\right) + l\ cos\left(x_7\right)\right)^2\right) + \frac{\mu}{4\, x_1^{\,3}}\ \left(J_x + J_y + J_z - 3\ \left(J_z\right.\right.$

$\qquad \left.\left. + \left(J_y - J_x\right)\ cos\left(2\ x_5\right)\right)\right) + \frac{1}{2}\, \frac{EA}{l_0} x_9^{\,2}$

$U := -\frac{\mu M}{x_1} + \frac{1}{2}\, \frac{\mu m_2\left(a^2 + l^2 + 2\, a\, l\cos\left(x_5 - x_7\right) - 3\left(a\cos\left(x_5\right) + l\cos\left(x_7\right)\right)^2\right)}{x_1^3}$ (2)

$\qquad + \frac{1}{4}\, \frac{\mu\left(J_x + J_y - 2\, J_z - 3\left(J_y - J_x\right)\cos\left(2\, x_5\right)\right)}{x_1^3} + \frac{1}{2}\, \frac{EA\, x_9^2}{l_0}$

> Lagrange := $T - U$

$Lagrange := \frac{1}{2}\, M\left(x_2^2 + x_4^2 x_1^2\right) + \frac{1}{2}\, m_2\left(w^2 a^2 + x_{10}^2 + w_l^2\, l^2 + 2\, a\, w\left(-x_{10}\sin\left(x_5 - x_7\right)\right.\right.$ (3)

$\qquad \left.\left. + w_l\, l\cos\left(x_5 - x_7\right)\right)\right) + \frac{1}{2}\, J_z\, w^2$

$\qquad + \frac{\mu M}{x_1} - \frac{1}{2}\, \frac{\mu m_2\left(a^2 + l^2 + 2\, a\, l\cos\left(x_5 - x_7\right) - 3\left(a\cos\left(x_5\right) + l\cos\left(x_7\right)\right)^2\right)}{x_1^3}$

$\qquad - \frac{1}{4}\, \frac{\mu\left(J_x + J_y - 2\, J_z - 3\left(J_y - J_x\right)\cos\left(2\, x_5\right)\right)}{x_1^3} - \frac{1}{2}\, \frac{EA\, x_9^2}{l_0}$

> Lagrange1 := $subs\left(\left\{w = x_4 + x_6,\ w_l = x_4 + x_8\right\},\ Lagrange\right)$

$Lagrange1 := \frac{1}{2}\, M\left(x_2^2 + x_4^2 x_1^2\right) + \frac{1}{2}\, m_2\left(\left(x_4 + x_6\right)^2 a^2 + x_{10}^2 + \left(x_4 + x_8\right)^2 l^2 + 2\, a\left(x_4\right.\right.$ (4)

$\qquad \left. + x_6\right)\left(-x_{10}\sin\left(x_5 - x_7\right) + \left(x_4 + x_8\right) l\cos\left(x_5 - x_7\right)\right)\Big) + \frac{1}{2}\, J_z\left(x_4 + x_6\right)^2$

$\qquad + \frac{\mu M}{x_1} \cdot \frac{1}{2}\, \frac{\mu m_2\left(a^2 + l^2 + 2\, a\, l\cos\left(x_5 - x_7\right) - 3\left(a\cos\left(x_5\right) + l\cos\left(x_7\right)\right)^2\right)}{x_1^3}$

$\qquad - \frac{1}{4}\, \frac{\mu\left(J_x + J_y - 2\, J_z - 3\left(J_y - J_x\right)\cos\left(2\, x_5\right)\right)}{x_1^3} - \frac{1}{2}\, \frac{EA\, x_9^2}{l_0}$

>

> $g_t := subs\Big($

Figure 6.2: Maple program for Problem 6.19 (contd.).

$\left\{x_1 = x_1(t), x_2 = x_2(t), x_3 = x_3(t), x_4 = x_4(t), x_5 = x_5(t), x_6 = x_6(t), x_7 = x_7(t), x_8 = x_8(t), x_9 = x_9\right.$
$\left.(t), x_{10} = x_{10}(t)\right\}, g\Big)$

$$g_t := M x_2(t) \tag{5}$$

$>\ Eq_1 := diff\big(g_t,\ t\big) - diff\big(Lagrange1,\ x_1\big) = 0$

$$Eq_1 := M\left(\frac{\mathrm{d}}{\mathrm{d}t}\, x_2(t)\right) - M x_4^2 x_1 \tag{6}$$

$$+ \frac{\mu M}{x_1^2} - \frac{3}{2}\, \frac{\mu m_2 \left(a^2 + l^2 + 2\, a\, l \cos\big(x_5 - x_7\big) - 3\, \big(a \cos\big(x_5\big) + l \cos\big(x_7\big)\big)^2\right)}{x_1^4}$$

$$- \frac{3}{4}\, \frac{\mu \left(J_x + J_y - 2\, J_z - 3\, \big(J_y - J_x\big)\cos\big(2\, x_5\big)\right)}{x_1^4} = 0$$

$>\ g := diff\big(Lagrange1,\ x_4\big)$

$$g := M x_4 x_1^2 + \frac{1}{2}\, m_2 \left(2\, \big(x_4 + x_6\big) a^2 + 2\, \big(x_4 + x_8\big) l^2 + 2\, a\, \big(-x_{10} \sin\big(x_5 - x_7\big) + \big(x_4 \right. \tag{7}$$

$$\left. + x_8\big) l \cos\big(x_5 - x_7\big)\big) + 2\, a\, \big(x_4 + x_6\big) l \cos\big(x_5 - x_7\big)\right) + J_z \big(x_4 + x_6\big)$$

$>\ g_t := subs\Big($

$\left\{x_1 = x_1(t), x_2 = x_2(t), x_3 = x_3(t), x_4 = x_4(t), x_5 = x_5(t), x_6 = x_6(t), x_7 = x_7(t), x_8 = x_8(t), x_9 = x_9\right.$
$\left.(t), x_{10} = x_{10}(t)\right\}, g\Big)$

$$g_t := M x_4(t) x_1(t)^2 + \frac{1}{2}\, m_2 \left(2\, \big(x_4(t) + x_6(t)\big) a^2 + 2\, \big(x_4(t) + x_8(t)\big) l^2 \right. \tag{8}$$

$$+ 2\, a\, \big(-x_{10}(t) \sin\big(x_5(t) - x_7(t)\big) + \big(x_4(t) + x_8(t)\big) l \cos\big(x_5(t) - x_7(t)\big)\big) + 2\, a\, \big(x_4(t)$$

$$\left. + x_6(t)\big) l \cos\big(x_5(t) - x_7(t)\big)\right) + J_z \big(x_4(t) + x_6(t)\big)$$

$>\ Eq_2 := diff\big(g_t,\ t\big) - diff\big(Lagrange1,\ x_3\big) = 0$

$$Eq_2 := M\left(\frac{\mathrm{d}}{\mathrm{d}t}\, x_4(t)\right) x_1(t)^2 + 2 M x_4(t) x_1(t) \left(\frac{\mathrm{d}}{\mathrm{d}t}\, x_1(t)\right) + \frac{1}{2}\, m_2 \left(2\left(\frac{\mathrm{d}}{\mathrm{d}t}\, x_4(t) + \frac{\mathrm{d}}{\mathrm{d}t}\, x_6(t)\right) a^2\right. \tag{9}$$

$$+ 2\left(\frac{\mathrm{d}}{\mathrm{d}t}\, x_4(t) + \frac{\mathrm{d}}{\mathrm{d}t}\, x_8(t)\right) l^2$$

$$+ 2\, a\, \left(-\left(\frac{\mathrm{d}}{\mathrm{d}t}\, x_{10}(t)\right) \sin\big(x_5(t) - x_7(t)\big)\right.$$

$$- x_{10}(t) \cos\big(x_5(t) - x_7(t)\big) \left(\frac{\mathrm{d}}{\mathrm{d}t}\, x_5(t) - \frac{\mathrm{d}}{\mathrm{d}t}\, x_7(t)\right) + \left(\frac{\mathrm{d}}{\mathrm{d}t}\, x_4(t)\right.$$

$$+ \frac{\mathrm{d}}{\mathrm{d}t}\, x_8(t)\right) l \cos\big(x_5(t) - x_7(t)\big) - \big(x_4(t)$$

$$+ x_8(t)\big) l \sin\big(x_5(t) - x_7(t)\big) \left(\frac{\mathrm{d}}{\mathrm{d}t}\, x_5(t) - \frac{\mathrm{d}}{\mathrm{d}t}\, x_7(t)\right)\right) + 2\, a\, \left(\frac{\mathrm{d}}{\mathrm{d}t}\, x_4(t)\right.$$

$$+ \frac{\mathrm{d}}{\mathrm{d}t}\, x_6(t)\right) l \cos\big(x_5(t) - x_7(t)\big) - 2\, a\, \big(x_4(t)$$

$$+ x_6(t)\big) l \sin\big(x_5(t) - x_7(t)\big) \left(\frac{\mathrm{d}}{\mathrm{d}t}\, x_5(t) - \frac{\mathrm{d}}{\mathrm{d}t}\, x_7(t)\right)\right) + J_z \left(\frac{\mathrm{d}}{\mathrm{d}t}\, x_4(t) + \frac{\mathrm{d}}{\mathrm{d}t}\, x_6(t)\right) = 0$$

$>\ g := diff\big(Lagrange1,\ x_6\big)$

Figure 6.3: Maple program for Problem 6.19 (contd.).

$$g := \frac{1}{2}\, m_2 \left(2\,(x_4 + x_6)\, a^2 + 2\, a\, \left(-x_{10} \sin(x_5 - x_7) + (x_4 + x_8)\, l \cos(x_5 - x_7) \right) \right) + J_z\, (x_4 + x_6) \qquad (10)$$

`> g_t := subs(`

$\{ x_1 = x_1(t),\, x_2 = x_2(t),\, x_3 = x_3(t),\, x_4 = x_4(t),\, x_5 = x_5(t),\, x_6 = x_6(t),\, x_7 = x_7(t),\, x_8 = x_8(t),\, x_9 = x_9$
$(t),\, x_{10} = x_{10}(t) \},\, g \big)$

$$g_t := \frac{1}{2}\, m_2 \left(2\,(x_4(t) + x_6(t))\, a^2 + 2\, a\, \left(-x_{10}(t) \sin(x_5(t) - x_7(t)) + (x_4(t) \right.\right. \qquad (11)$$
$$\left.\left. + x_8(t))\, l \cos(x_5(t) - x_7(t)) \right) \right) + J_z\, (x_4(t) + x_6(t))$$

`> Eq_3 := diff(g_t, t) - diff(Lagrange1, x_5) = 0`

$$Eq_3 := \frac{1}{2}\, m_2 \left(2 \left(\frac{d}{dt} x_4(t) + \frac{d}{dt} x_6(t) \right) a^2 \right. \qquad (12)$$
$$+ 2\, a \left(-\left(\frac{d}{dt} x_{10}(t) \right) \sin(x_5(t) - x_7(t)) \right.$$
$$- x_{10}(t) \cos(x_5(t) - x_7(t)) \left(\frac{d}{dt} x_5(t) - \frac{d}{dt} x_7(t) \right) + \left(\frac{d}{dt} x_4(t) \right.$$
$$+ \frac{d}{dt} x_8(t) \right) l \cos(x_5(t) - x_7(t)) - (x_4(t)$$
$$+ x_8(t)) l \sin(x_5(t) - x_7(t)) \left(\frac{d}{dt} x_5(t) - \frac{d}{dt} x_7(t) \right) \right) + J_z \left(\frac{d}{dt} x_4(t) \right.$$
$$+ \frac{d}{dt} x_6(t) \right) - m_2\, a\, (x_4 + x_6) \left(-x_{10} \cos(x_5 - x_7) - (x_4 + x_8)\, l \sin(x_5 - x_7) \right)$$
$$+ \frac{1}{2} \frac{\mu\, m_2 \left(-2\, a\, l \sin(x_5 - x_7) + 6\, (a \cos(x_5) + l \cos(x_7))\, a \sin(x_5) \right)}{x_1^3}$$
$$+ \frac{3}{2} \frac{\mu\, (J_y - J_x) \sin(2 x_5)}{x_1^3} = 0$$

`> g := diff(Lagrange1, x_8)`

$$g := \frac{1}{2}\, m_2 \left(2\,(x_4 + x_8)\, l^2 + 2\, a\, (x_4 + x_6)\, l \cos(x_5 - x_7) \right) \qquad (13)$$

`> g_t := subs(`

$\{ x_1 = x_1(t),\, x_2 = x_2(t),\, x_3 = x_3(t),\, x_4 = x_4(t),\, x_5 = x_5(t),\, x_6 = x_6(t),\, x_7 = x_7(t),\, x_8 = x_8(t),\, x_9 = x_9$
$(t),\, x_{10} = x_{10}(t) \},\, g \big)$

$$g_t := \frac{1}{2}\, m_2 \left(2\,(x_4(t) + x_8(t))\, l^2 + 2\, a\, (x_4(t) + x_6(t))\, l \cos(x_5(t) - x_7(t)) \right) \qquad (14)$$

`> Eq_4 := diff(g_t, t) - diff(Lagrange1, x_7) = 0`

$$Eq_4 := \frac{1}{2}\, m_2 \left(2 \left(\frac{d}{dt} x_4(t) + \frac{d}{dt} x_8(t) \right) l^2 + 2\, a \left(\frac{d}{dt} x_4(t) \right.\right. \qquad (15)$$
$$+ \frac{d}{dt} x_6(t) \right) l \cos(x_5(t) - x_7(t)) - 2\, a\, (x_4(t)$$
$$+ x_6(t)) l \sin(x_5(t) - x_7(t)) \left(\frac{d}{dt} x_5(t) - \frac{d}{dt} x_7(t) \right) \right) - m_2\, a\, (x_4 + x_6) \left(x_{10} \cos(x_5 - x_7) \right.$$

Figure 6.4: Maple program for Problem 6.19 (contd.).

$$+ \left(x_4 + x_8 \right) l \sin \left(x_5 - x_7 \right)$$

$$+ \frac{1}{2} \frac{\mu \, m_2 \left(2 \, a \, l \sin \left(x_5 - x_7 \right) + 6 \left(a \cos \left(x_5 \right) + l \cos \left(x_7 \right) \right) l \sin \left(x_7 \right) \right)}{x_1^3} = 0$$

\> $g := \mathit{diff}\left(\mathit{Lagrange1}, x_{10} \right)$

$$g := \frac{1}{2} \, m_2 \left(2 \, x_{10} - 2 \, a \left(x_4 + x_6 \right) \sin \left(x_5 - x_7 \right) \right) \tag{16}$$

\> $g_t := \mathit{subs}\bigg($

$\Big\{ x_1 = x_1(t), x_2 = x_2(t), x_3 = x_3(t), x_4 = x_4(t), x_5 = x_5(t), x_6 = x_6(t), x_7 = x_7(t), x_8 = x_8(t), x_9 = x_9$
$(t), x_{10} = x_{10}(t) \Big\}, g \bigg)$

$$g_t := \frac{1}{2} \, m_2 \left(2 \, x_{10}(t) - 2 \, a \left(x_4(t) + x_6(t) \right) \sin \left(x_5(t) - x_7(t) \right) \right) \tag{17}$$

\> $\mathit{Eq}_4 := \mathit{diff}\left(g_t, t \right) - \mathit{diff}\left(\mathit{Lagrange1}, x_9 \right) = 0$

$$\mathit{Eq}_4 := \frac{1}{2} \, m_2 \left(2 \left(\frac{\mathrm{d}}{\mathrm{d}t} x_{10}(t) \right) - 2 \, a \left(\frac{\mathrm{d}}{\mathrm{d}t} x_4(t) + \frac{\mathrm{d}}{\mathrm{d}t} x_6(t) \right) \sin \left(x_5(t) - x_7(t) \right) - 2 \, a \left(x_4(t) \right. \tag{18}$$

$$\left. + x_6(t) \right) \cos \left(x_5(t) - x_7(t) \right) \left(\frac{\mathrm{d}}{\mathrm{d}t} x_5(t) - \frac{\mathrm{d}}{\mathrm{d}t} x_7(t) \right) \bigg) + \frac{EA \, x_9}{l_0} = 0$$

Figure 6.5: Maple program for Problem 6.19.

Chapter 7

Control System

Problem Set 7

7.1

(a) The system is defined by the equation

$$I\ddot{\theta} = T \tag{7.1}$$

Applying Laplace Transform, we get

$$I\left(s^2\theta(s) - s\theta_0 - \dot{\theta}_0\right) = T(s) \tag{7.2}$$

Rewriting the preceding equation with output as $\theta(s)$,

$$\theta(s) = \frac{1}{Is^2}T(s) + \frac{1}{s}\theta_0 + \frac{1}{s^2}\dot{\theta}_0 \tag{7.3}$$

Thus, the open-loop transfer function of the given system with input as $T(s)$ and output as $\theta(s)$ is

$$G(s) = \frac{\theta(s)}{T(s)} = \frac{1}{Is^2} \tag{7.4}$$

The characteristic equation of the open-loop system is

$$Is^2 = 0 \qquad (7.5)$$

The roots of the characteristic equation are

$$s_{1,2} = 0 \qquad (7.6)$$

This open-loop system is neutral stable (*i.e.,* θ remains constant) if input T=0 and $\dot{\theta}_0$=0 as the system response is $\theta(t) = \theta_0 + \dot{\theta}_0 t = \theta_0$. However, with the step input of T, the system would be unstable as the roots are zero. (**Note.** Please find the solution in this case, and verify the system stability.)

(b) Referring to the system defined by Eq. (7.1), the corresponding solution with initial attitude angle θ_0 and rate $\dot{\theta}_0$ is

$$\theta(t) = \theta_0 + \dot{\theta}_0 t \qquad (7.7)$$

(c) As the gas jet provides a constant thrust, the input is a step input.

(d)

(d.1) The characteristic equation of the closed loop system with a proportional controller is

$$1 + G_c(s)G(s) = 0 \qquad (7.8)$$

where $G_c(s)$ is the controller transfer function given as $G_c(s) = k_p$. The term k_p is the proportional gain.

Substituting $G(s)$ from Eq.(7.4) into Eq. (7.8), we get

$$Is^2 + k_p = 0 \qquad (7.9)$$

The roots of the characteristic Eq. (7.9) are

$$s_{1,2} = \pm j\sqrt{\frac{k_p}{I}} = \pm j\sqrt{\frac{k_p}{100}} \qquad (7.10)$$

As the roots of the characteristic equation of the closed loop system lie on the imaginary axis, the closed loop system is marginally stable with a proportional controller.

For the system stability, the roots in Eq. (7.10) should not have positive real parts, *i.e.,*

$$k_p > 0 \quad \Rightarrow k_p \in (0, \infty) \tag{7.11}$$

The system equation of motion with a proportional controller is given by

$$I\ddot{\theta} = k_p(\theta_r - \theta) \tag{7.12}$$

or

$$I\ddot{\theta} + k_p\theta = k_p\theta_r \tag{7.13}$$

where θ_r is the reference angle.

The solution of Eq. (7.13) with the initial attitude error θ_0 is

$$\theta(t) = \theta_r + (\theta_0 - \theta_r)\cos\sqrt{\frac{k_p}{I}}t \tag{7.14}$$

(d.2) The *final value of theorem* of Laplace Transform can not be applied here to find the steady state error as the system is not asymptotically stable.

(d.3) The characteristic equation of the closed loop system with a derivative controller is

$$1 + G_c(s)G(s) = 0 \tag{7.15}$$

where $G_c(s)$ is the controller transfer function given as $G_c(s) = sk_d$. The term k_d is the derivative gain.

Substituting $G(s)$ from Eq.(7.4) into Eq. (7.15), we get

$$Is^2 + k_d s = 0 \tag{7.16}$$

The roots of the characteristic Eq. (7.16) are

$$s_1 = 0, \quad s_2 = -\frac{k_d}{I} = 0 \tag{7.17}$$

This open-loop system is asymptotically stable (*i.e.*, θ diminishes and reaches to steady state value) if input T=0. With the step input of T, the system would be still remain stable as the pole $s_1 = 0$ gets cancelled out with the zero $s = 0$ at numerator.

For the system stability, the roots in Eq. (7.17) should not have positive real parts, *i.e.*,

$$k_d > 0 \quad \Rightarrow k_d \in (0, \infty) \tag{7.18}$$

The system equation of motion with a derivative controller is given by

$$I\ddot{\theta} = k_d(\dot{\theta}_r - \dot{\theta}) \tag{7.19}$$

or

$$I\ddot{\theta} + k_d\dot{\theta} = k_d\dot{\theta}_r \tag{7.20}$$

Applying Laplace Transform with initial conditions as null, we get

$$(Is^2\theta(s) + k_d s\theta(s)) = k_d s\theta_r(s) \tag{7.21}$$

Rewriting the preceding equation with output as $\theta(s)$

$$\theta(s) = \frac{k_d s\theta_r(s)}{Is^2 + k_d s} \tag{7.22}$$

Taking Laplace inverse transform,

$$\theta(t) = L^{-1}\left[\frac{k_d s\theta_r(s)}{Is^2 + k_d s}\right] = L^{-1}s\theta_r(s)\left[\frac{1}{s} - \frac{1}{s + \frac{k_d}{I}}\right] \tag{7.23}$$

The solution of Eq. (7.23) assuming a step input $\theta_r(s)$ is

$$\theta(t) = \theta_r\left(1 - e^{-\frac{k_d}{I}t}\right) \tag{7.24}$$

The *final value of theorem* of Laplace Transform can be applied here to find the steady state error as the system is asymptotically stable.

The steady state value is

$$\theta(\infty) = \lim_{t \to \infty} \theta_r [1 - e^{\frac{k_d}{I} t}] = \theta_r \qquad (7.25)$$

Thus, the steady state error is

$$e_{ss} = \theta_r - \theta(\infty) = \theta_r - \theta_r = 0 \qquad (7.26)$$

(d.4) The characteristic equation of the closed loop system with a proportional-derivative (PD) controller is

$$1 + G_c(s)G(s) = 0 \qquad (7.27)$$

where $G_c(s)$ is the controller transfer function given as $G_c(s) = k_p + sk_d$. The terms k_p and k_d are the proportional and derivative gains.

Substituting $G(s)$ from Eq.(7.4) into Eq. (7.27), we get

$$Is^2 + k_d s + k_p = 0 \qquad (7.28)$$

The roots of the characteristic Eq. (7.16) are

$$s_{1,2} = \frac{-k_d \pm \sqrt{k_d^2 - 4k_p I}}{2I} \qquad (7.29)$$

Apply the Routh-Hurwitz criterion to the characteristic Eq. (7.16), we get the stability conditions as

$$\begin{aligned} k_p > 0 &\quad \Rightarrow k_p \in (0, \infty) \\ k_d > 0 &\quad \Rightarrow k_d \in (0, \infty) \end{aligned} \qquad (7.30)$$

(d.5) The system equation of motion with a proportional-derivative controller is given as

$$I\ddot{\theta} = k_p(\theta_r - \theta) + k_d(\dot{\theta}_r - \dot{\theta}) \qquad (7.31)$$

or

$$I\ddot{\theta} + k_d\dot{\theta} + k_p\theta = k_p\theta_r + k_d\dot{\theta}_r \tag{7.32}$$

The corresponding characteristic equation of the given system is

$$Is^2 + k_ds + k_p = 0 \tag{7.33}$$

Rewriting Eq. (7.33)

$$s^2 + \frac{k_d}{I}s + \frac{k_p}{I} = 0 \tag{7.34}$$

Comparing the preceding characteristic Eq. (7.34) with the characteristic equation of a standard second-order system (where ω_n is the undamped natural frequency and ζ is the damping ratio):

$$s^2 + 2\zeta\omega_n s + \omega_n^2 = 0 \tag{7.35}$$

we get

$$\frac{k_d}{I} = 2\zeta\omega_n \tag{7.36}$$

$$\frac{k_p}{I} = \omega_n^2 \tag{7.37}$$

The natural frequency ω_n is

$$\omega_n = \frac{2\pi f}{\sqrt{1 - \zeta^2}}$$

So,

$$k_d = 2I\zeta\omega_n = 2 \times 100 \times 0.5 \times \frac{2\pi \times 0.25}{\sqrt{1 - 0.5^2}} = 181.4 \tag{7.38}$$

$$k_p = I\omega_n^2 = 100 \times \left(\frac{2\pi \times 0.25}{\sqrt{1 - 0.5^2}}\right)^2 = 284.91 \tag{7.39}$$

7.2 *Given.*

$\dot{\theta}_r = 3^0/\text{s}$; $I = 20$ kg-m^2; $c = 5$ N-sec/m;

$e_{ss} \leq 0.1^o$; $\tau \leq 0.5$ sec

The system equation of motion with a proportional-derivative controller is

$$I\ddot{\theta} + c\dot{\theta} = k_p(\theta_r - \theta) + k_d(\dot{\theta}_r - \dot{\theta}) \qquad (7.40)$$

or

$$I\ddot{\theta} + (c + k_d)\dot{\theta} + k_p\theta = k_p\theta_r + k_d\dot{\theta}_r \qquad (7.41)$$

The corresponding characteristic equation of the given system is

$$Is^2 + (c + k_d)s + k_p = 0 \qquad (7.42)$$

Rewriting Eq. (7.42)

$$s^2 + \frac{c + k_d}{I}s + \frac{k_p}{I} = 0 \qquad (7.43)$$

Comparing the preceding characteristic Eq. (7.43) with the characteristic equation of a standard second-order system (where ω_n is the undamped natural frequency and ζ is the damping ratio):

$$s^2 + 2\zeta\omega_n s + \omega_n^2 = 0 \qquad (7.44)$$

we get

$$\frac{c + k_d}{I} = 2\zeta\omega_n \qquad (7.45)$$

$$\frac{k_p}{I} = \omega_n^2 \qquad (7.46)$$

The time constant τ is

$$\tau = \frac{1}{\zeta\omega_n} \qquad (7.47)$$

Substituting for the term $\zeta\omega_n$ from Eq. (7.46) into Eq. (7.47),

$$\tau = \frac{2I}{c + k_d} \le 0.5$$

or

$$\frac{c + k_d}{2I} \ge 2 \quad \Rightarrow \quad \frac{5 + k_d}{2 \times 20} \ge 2$$

Solving we get

$$k_d \ge 75 \tag{7.48}$$

The closed-loop transfer function of the system given by Eq. (7.41) can be expressed as

$$M(s) = \frac{\theta(s)}{\theta_r(s)} = \frac{k_p + sk_d}{Is^2 + (c + k_d)s + k_p} \tag{7.49}$$

The steady-state error is

$$e_{ss} = \lim_{s \to 0} sE(s) = \lim_{s \to 0} s[\theta_r(s) - \theta(s)] = \lim_{s \to 0} s\theta_r(s)[1 - M(s)] \tag{7.50}$$

Here the input is ramp input. Substituting $M(s)$ from Eq. (7.49) into the preceding Eq. (7.50), we have

$$e_{ss} = \lim_{s \to 0} s\frac{\dot{\theta}_r}{s^2}\frac{Is^2 + cs}{Is^2 + (c + k_d)s + k_p} \tag{7.51}$$

$$= \lim_{s \to 0} \dot{\theta}_r \frac{Is + c}{Is^2 + (c + k_d)s + k_p} = \dot{\theta}_r \frac{c}{k_p} \tag{7.52}$$

Thus,

$$\dot{\theta}_r \frac{c}{k_p} \le 0.1\text{deg}$$

$$3\text{deg/sec}\frac{5\text{N-m-sec/rad}}{k_p} \le 0.1\text{deg}$$

or

$$k_p \ge \frac{15}{0.1} = 150\text{N-m/rad} \tag{7.53}$$

(c) No, the desired performance can not be met.

7.3 *Given.*

$\dot{\theta}_r = 3^0/\text{s}$;

m=20 kg; k=1 N/s; c=4 N-sec/m;

$t_s \leq 1sec$; $M_p \leq 0.2$

Considering settling time (to within 5% of the final value),

$$t_s = \frac{3}{\zeta\omega_n} \leq 1 \tag{7.54}$$

$$\frac{3 \times 2}{c + k_d} \leq 1 \quad \Rightarrow 6 \leq 4 + k_d \tag{7.55}$$

or

$$k_d \geq 2 \tag{7.56}$$

If we take settling time (to within 1% of the final value),

$$t_s = \frac{4.6}{\zeta\omega_n} \leq 1 \tag{7.57}$$

$$\frac{4.6 \times 2}{c + k_d} \leq 1 \quad \Rightarrow 9.2 \leq 4 + k_d \tag{7.58}$$

or

$$k_d \geq 5.2 \tag{7.59}$$

For maximum overshoot,

$$\begin{aligned}\zeta &\geq \sqrt{\frac{(lnM_p/\pi)^2}{1 + (lnM_p/\pi)^2}} \\ &\geq \sqrt{\frac{(ln0.2/\pi)^2}{1 + (ln0.2/\pi)^2}} = 0.4559\end{aligned} \tag{7.60}$$

or

$$\frac{c + k_d}{2m\omega_n} \geq 0.4559 \quad \Rightarrow \frac{c + k_d}{2\sqrt{(k + k_p)m}} \geq 0.4559$$

or

$$\frac{4 + k_d}{2 \times 0.4559} \geq \sqrt{1 + k_p} \quad \Rightarrow k_p \leq 1.2025(4 + k_d)^2 - 1 \tag{7.61}$$

7.4 The pitch equation of motion of the spacecraft is

$$I\ddot{\theta} + c\dot{\theta} = T_d \tag{7.62}$$

Applying Laplace Transform with initial conditions $\theta = \theta_0 = 0$ and $\theta = \dot{\theta}_0$, we get

$$I(s\dot{\theta}(s) - \dot{\theta}_0) + c\dot{\theta}(s) = 1 \tag{7.63}$$

$$I(s^2\theta(s) - \dot{\theta}_0) + cs\theta(s) = 1 \tag{7.64}$$

$$\theta(s) = \frac{1 + I\dot{\theta}_0}{Is^2 + cs} \tag{7.65}$$

$$\theta(s) = \frac{1 + I\dot{\theta}_0}{Is^2 + cs} \tag{7.66}$$

Taking inverse Laplace transform,

$$\theta(t) = L^{-1}\left[\frac{1 + I\dot{\theta}_0}{Is^2 + cs}\right] = L^{-1}\frac{1 + I\dot{\theta}_0}{c}\left[\frac{1}{s} - \frac{1}{s + \frac{c}{I}}\right] \tag{7.67}$$

The solution of Eq. (7.67) is

$$\theta(t) = \frac{1 + I\dot{\theta}_0}{c}(1 - e^{-\frac{c}{I}t}) \tag{7.68}$$

or

$$\theta(t) = \frac{1 + 100\dot{\theta}_0}{10}(1 - e^{-\frac{10}{100}t})$$
$$= (0.1 + 10\dot{\theta}_0)(1 - e^{-0.1t}) \tag{7.69}$$

So, differentiating Eq. (7.69), we obtain $\dot{\theta}(t)$ as

$$\dot{\theta}(t) = (0.01 + \dot{\theta}_0)e^{-0.1t} \tag{7.70}$$

The final value of $\theta(t)$ as t tends to ∞ is

$$\theta(t = \infty) = \lim_{t \to \infty} (0.1 + 10\dot{\theta}_0)(1 - e^{-0.1t}) = 0.1 + 10\dot{\theta}_0 \qquad (7.71)$$

The open-loop transfer function of the system

$$G(s) = \frac{1}{Is^2 + cs} \qquad (7.72)$$

The characteristic equation of the uncontrolled system,

$$Is^2 + cs = 0 \qquad (7.73)$$

Thus, the eigenvalues are

$$s = 0; \quad s = \frac{c}{I} = \frac{10}{100} = 0.1 \qquad (7.74)$$

Applying a proportional controller, the characteristic equation of the controlled system is

$$1 + G_c(s)G(s) = 0 \qquad (7.75)$$

where $G_c(s)$ is the controller transfer function given as $G_c(s) = k_p$. The term k_p is the proportional gain.

Thus, the characteristic equation of the controlled system

$$Is^2 + cs + k_p = 0 \qquad (7.76)$$

The eigenvalues are

$$s_{1,2} = \frac{-c \pm \sqrt{c^2 - 4Ik_p}}{2I} = \frac{-10 \pm \sqrt{100 - 4 \times 100k_p}}{200} \qquad (7.77)$$

For $k_p = 50$, the eigenvalues are

$$s_{1,2} = \frac{-10 \pm \sqrt{100 - 4 \times 100 \times 50}}{200} = -0.05 \pm j0.7053 \qquad (7.78)$$

The closed-loop system will be asymptotically stable.

7.5 **(a)** The open-loop transfer function of the system is

$$G(s) = \frac{\theta(s)}{\delta_e(s)} = \frac{1.151s + 0.1774}{s^3 + 0.739s^2 + 0.921s} \qquad (7.79)$$

The characteristic equation is

$$s^3 + 0.739s^2 + 0.921s = 0 \qquad (7.80)$$

The roots of the characteristic equation are

$$s_1 = 0, \quad s_{1,2} = -0.3695 \pm j0.866 \qquad (7.81)$$

In the case, the input is zero. The system response is asymptotically stable with response as

$$\theta = c_1 + c_2 e^{-0.3695t} sin(0.866t + c_3) \qquad (7.82)$$

where $c_j, j = 1, 2, 3$ are constants depending upon the initial conditions.

When the step input of δ_e is applied, the steady-state θ $(t = \infty)$, applying final value theorem, is

$$\theta(t \to \infty) = \lim_{s \to 0} s\theta(s) = \lim_{s \to 0} sG(s)\delta_e(s)$$

$$= \lim_{s \to 0} s \left(\frac{1.151s + 0.1774}{s^3 + 0.739s^2 + 0.921s} \right) \frac{\theta_r}{s} = \infty \qquad (7.83)$$

Thus, the system is unstable. In fact, in case one of the roots (*i.e.*, s_1) is zero, and we apply step input, the system would be unstable if this pole does not get cancelled out by zero at the numerator.

(b) Applying proportional control, the closed-loop transfer function of the system is

$$M(s) = \frac{(k_p)G(s)}{1 + k_pG(s)}$$

$$= \frac{1.151k_ps + 0.1774k_p}{s^3 + 0.739s^2 + (0.921 + 1.151k_p)s + 0.1774k_p} \qquad (7.84)$$

The characteristic equation is

$$s^3 + 0.739s^2 + (0.921 + 1.151k_p)s + 0.1774k_p = 0 \qquad (7.85)$$

The Routh array is as follows

$$
\begin{array}{c c c}
s^3 & 1 & 0.921 + 1.151k_p \\[4pt]
s^2 & 0.739 & 0.1774k_p \\[4pt]
s^1 & 0.921 + 1.151k_p - \frac{0.1774k_p}{0.739} & 0 \\[4pt]
s^0 & 0.1774k_p &
\end{array}
\qquad (7.86)
$$

Applying the Routh-Hurwitz conditions of absolute stability,

(a) All coefficients in the characteristic equations should be positive, *i.e.*,

$$0.921 + 1.151k_p > 0, \quad 0.1774k_p > 0$$

(b) All elements in the first column of the Routh array should be positive, *i.e.*,

$$0.921 + 1.151k_p - \frac{0.1774k_p}{0.739}, \quad 0.1774k_p > 0$$

Thus, we can write the conditions of the system stability as

$$k_p > -1.011 \text{ and } k_p > 0 \quad \Rightarrow k_p > 0 \qquad (7.87)$$

The final value of X for a unit step input is

$$
\begin{aligned}
X(\infty) &= \lim_{s \to 0} sM(s)F(s) \\
&= \lim_{s \to 0} s \left[\frac{1.151k_p s + 0.1774k_p}{s^3 + 0.739s^2 + (0.921 + 1.151k_p)s + 0.1774k_p} \right] \frac{1}{s} \\
&= 1 \qquad\qquad\qquad\qquad\qquad\qquad\qquad\qquad\qquad\qquad (7.88)
\end{aligned}
$$

The steady state error e_{ss} is

$$e_{ss} = 1 - X(\infty) = 1 - 1 = 0 \qquad (7.89)$$

(c) Applying proportional-plus-derivative control, the closed-loop transfer function of the system is

$$M(s) = \frac{(k_p + k_d s)G(s)}{1 + (k_p + k_d s)G(s)}$$

$$= \frac{1.151k_d s^2 + (1.151k_p + 0.1774k_d)s + 0.1774k_p}{s^3 + (0.739 + 1.151k_d)s^2 + (0.921 + 1.151k_p + 0.1774k_d)s + 0.1774k_p}$$

$$(7.90)$$

The characteristic equation is

$$s^3 + (0.739 + 1.151k_d)s^2 + (0.921 + 1.151k_p + 0.1774k_d)s + 0.1774k_p = 0$$

$$(7.91)$$

The Routh array is as follows

s^3	1	$0.921 + 1.151k_p + 0.1774k_d$
s^2	$0.739 + 1.151k_d$	$0.1774k_p$
s^1	$0.921 + 1.151k_p + 0.1774k_d - \frac{0.1774k_p}{0.739}$	0
s^0	$0.1774k_p$	

$$(7.92)$$

Applying the Routh-Hurwitz conditions of absolute stability,

(a) All coefficients in the characteristic equations should be positive, *i.e.*,

$$0.921 + 1.151k_p > 0, \quad 0.1774k_p > 0$$

(b) All elements in the first column of the Routh array should be positive, *i.e.*,

$$0.921 + 1.151k_p - \frac{0.1774k_p}{0.739}, \quad 0.1774k_p > 0$$

Thus, we can write the conditions of the system stability as

$$k_p > -1.011 \text{ and } k_p > 0 \quad \Rightarrow k_p > 0 \qquad (7.93)$$

For $K_p{=}9$ and $K_d{=}8$, the characteristic equation is

$$s^3 + 9.947s^2 + 12.7s + 1.597 = 0 \qquad (7.94)$$

The Routh array is as follows

$$
\begin{array}{lll}
s^3 & 1 & 12.70 \\
s^2 & 0.947 & 1.597 \\
s^1 & 11.47 & 0 \\
s^0 & 1.597 &
\end{array}
\qquad (7.95)
$$

Applying the Routh-Hurwitz conditions of absolute stability, we can say the system is stable.

(d) Applying proportional-integral-derivative control, the closed-loop transfer function of the system is

$$
\begin{aligned}
M(s) &= \frac{(k_p + \frac{k_i}{s} + k_d s)G(s)}{1 + (k_p + + \frac{k_i}{s} + k_d s)G(s)} \\
&= \frac{1.151k_d s^2 + (1.151k_p + 0.1774k_d)s + 0.1774k_p}{s^3 + (0.739 + 1.151k_d)s^2 + (0.921 + 1.151k_p + 0.1774k_d)s + 0.1774k_p}
\end{aligned}
$$
$$(7.96)$$

For K_p=2, K_i=4, and K_d=3, the characteristic equation is

$$s^4 + 4.192s^3 + 3.755s^2 + 4.959s + 0.710 = 0 \qquad (7.97)$$

The Routh array is as follows

$$
\begin{array}{llll}
s^4 & 1 & 3.76 & 0.71 \\
s^3 & 4.19 & 4.96 & 0 \\
s^2 & 2.58 & 0.71 & 0 \\
s^1 & 3.81 & 0 & \\
s^0 & 0.71 & &
\end{array}
\qquad (7.98)
$$

Applying the Routh-Hurwitz conditions of absolute stability, we can say that the system is stable.

4.6 (a)

$$M_p = e^{-\dfrac{\pi\zeta}{\sqrt{1-\zeta^2}}} = 0 \quad \Rightarrow \zeta = 1 \qquad (7.99)$$

$$t_r = \frac{1.8}{\omega_n} \le 0.8 \quad \Rightarrow \omega_n \ge \frac{1.8}{0.8} = 2.25 \qquad (7.100)$$

Thus, the characteristic equation should be

$$s^2 + 2\zeta\omega_n s + \omega^2 = 0$$

or

$$s^2 + 2 \times 2.25s + 2.25^2 = 0 \Rightarrow s^2 + 4.5s + 5.0625 = 0$$

We place the last pole far enough from the first and the second poles which determine the response of the system. Let us take s_3=-20. Then, the desired characteristic equation is

$$(s + 20)(s^2 + 4.5s + 5.06) = s^3 + 24.5s^2 + 95.06s + 101.25 = 0$$

Applying proportional-integral-derivative (PID) control, the closed-loop transfer function of the system is

$$
\begin{aligned}
M(s) &= \frac{[k_p + \frac{k_i}{s} + sk_d]G(s)}{1 + [k_p + \frac{k_i}{s} + sk_d]G(s)} \\
&= \frac{k_i + k_p s + k_d s^2}{s^3 + (10 + k_d)s^2 + (20 + k_p)s + k_i} \qquad (7.101)
\end{aligned}
$$

The characteristic equation is

$$s^3 + (10 + k_d)s^2 + (20 + k_p)s + k_i = 0 \qquad (7.102)$$

Comparing the two equations,

$$k_i = 101.25, \ k_p = 75.06, \ k_d = 14.5$$

(b) The open-loop transfer function of the system is

$$G(s) = \frac{X(s)}{F(s)} = \frac{1}{s^2 + 10s + 20} \qquad (7.103)$$

The characteristic equation of the open-loop system is

$$s^2 + 10s + 20 = 0 \qquad (7.104)$$

Thus, the roots of the characteristic equation are

$$s_1 = -4.2361, \quad s_2 = -2.7639 \qquad (7.105)$$

As the two roots are in the left half s-plane, the open-loop system is stable.

The final value of X for the unit step input is

$$X(\infty) = \lim_{s \to 0} sG(s)F(s) = \lim_{s \to 0} s \times \frac{1}{s} \frac{1}{s^2 + 10s + 20}$$
$$= \frac{1}{20} \qquad (7.106)$$

The steady-state error e_{ss} is

$$e_{ss} = 1 - X(\infty) = 1 - \frac{1}{20} = \frac{19}{20} = 0.95 \qquad (7.107)$$

As e_{ss} is quite large, we need to apply a suitable controller.

(c) Applying proportional control, the closed-loop transfer function of the system is

$$M(s) = \frac{k_p G(s)}{1 + k_p G(s)} = \frac{k_p}{s^2 + 10s + 20 + k_p} \qquad (7.108)$$

The characteristic equation is

$$s^2 + 10s + 20 + k_p = 0 \qquad (7.109)$$

Applying the Routh-Hurwitz criterion, the stability condition is

$$k_p > -20 \quad \Rightarrow k_p \in (20, \infty) \qquad (7.110)$$

(d) Applying proportional-plus-derivative control, the closed-loop transfer function of the system is

$$M(s) = \frac{(k_p + sk_d)G(s)}{1 + (k_p + sk_d)G(s)} = \frac{k_p + sk_d}{s^2 + (10 + k_d)s + 20 + k_p}$$

$$(7.111)$$

The characteristic equation is

$$s^2 + (10 + k_d)s + 20 + k_p = 0 \qquad (7.112)$$

The Routh array is as follows

$$
\begin{array}{ccc}
s^2 & 1 & 20 + k_p \\
s^1 & 10 + k_d & 0 \\
s^0 & 20 + k_p &
\end{array}
\qquad (7.113)
$$

Applying the Routh-Hurwitz conditions of absolute stability,

(a) All coefficients in the characteristic equations should be positive, *i.e.*, $10 + k_d > 0, \quad 20 + k_p > 0$

(b) All elements in the first column of the Routh array should be positive, *i.e.*, $10 + k_d > 0, \quad 20 + k_p$

Thus, we can write the conditions of the system stability as

$$
\begin{aligned}
k_p &> -20 \\
k_d &> -10
\end{aligned}
\qquad (7.114)
$$

The final value of X for a unit step input is

$$X(\infty) = \lim_{s \to 0} sM(s)F(s) = \lim_{s \to 0} s \left[\frac{k_p + k_d s}{s^2 + (10 + k_d)s + 20 + k_p} \right] \frac{1}{s}$$

$$= \frac{k_p}{20 + k_p} \qquad (7.115)$$

The steady state error e_{ss} is

$$e_{ss} = 1 - X(\infty) = 1 - \frac{k_p}{20 + k_p} = \frac{20}{20 + k_p} \qquad (7.116)$$

If k_p is very large, then e_{ss} would be zero. Thus, with proportional-plus-derivative control the steady-state error would be present. For the given gains k_p=300 and k_d=10, the characteristic equation is

$$s^2 + 20s + 320 = 0 \qquad (7.117)$$

The characteristic roots are

$$s_{1,2} = -10 \pm \sqrt{\frac{400}{4} - 320} = -10 \pm 14.832 \qquad (7.118)$$

As all the roots are in the left half s-plane, the closed-loop system would be stable.

(e) Applying proportional-integral-derivative (PID) control, the closed-loop transfer function of the system is

$$M(s) = \frac{[k_p + \frac{k_i}{s} + sk_d]G(s)}{1 + [k_p + \frac{k_i}{s} + sk_d]G(s)}$$

$$= \frac{k_i + k_p s + k_d s^2}{s^3 + (10 + k_d)s^2 + (20 + k_p)s + k_i} \qquad (7.119)$$

The characteristic equation is

$$s^3 + (10 + k_d)s^2 + (20 + k_p)s + k_i = 0 \qquad (7.120)$$

The Routh array is as follows

$$
\begin{array}{ccc}
s^3 & 1 & 20 + k_p \\
s^2 & 10 + k_d & k_i \\
s^1 & 20 + k_p - \frac{k_i}{10+k_d} & 0 \\
s^0 & k_i &
\end{array}
\qquad (7.121)
$$

Applying the Routh-Hurwitz conditions of absolute stability,

(a) All coefficients in the characteristic equations should be positive, *i.e.*,

$$10 + k_d > 0, \quad 20 + k_p > 0, \quad k_i > 0$$

(b) All elements in the first column of the Routh array should be
positive, *i.e.,*

$$10 + k_d > 0, \quad 20 + k_p - \frac{k_i}{10 + k_d} > 0, \quad k_i > 0$$

Thus, we can write the conditions of the system stability as

$$k_p > \frac{k_i}{10 + k_d} - 20$$
$$k_i > 0 \tag{7.122}$$
$$k_d > -10$$

The final value of X for a unit step input is

$$X(\infty) = \lim_{s \to 0} sM(s)F(s) = \lim_{s \to 0} s \left[\frac{k_i + k_p s + k_d s^2}{s^3 + (10 + k_d)s^2 + (20 + k_p)s + k_i} \right] \frac{1}{s}$$
$$= \frac{k_i}{k_i} = 1 \tag{7.123}$$

The steady state error e_{ss} is

$$e_{ss} = 1 - X(\infty) = 1 - 1 = 0 \tag{7.124}$$

For the given gains k_p=350, k_i=300, and k_d=50, the characteristic
equation is

$$s^3 + 60s^2 + 370s + 300 = 0 \tag{7.125}$$

Using Matlab function "roots", we find the characteristic roots as

$$s_1 = -53.144, \ s_2 = -5.899, \ s_3 = -0.957 \tag{7.126}$$

As all the roots are in the left half s-plane, the closed-loop system
would be stable.

(f) Applying proportional-plus-integral (PI) control, the closed-loop
transfer function of the system is

$$M(s) = \frac{[k_p + \frac{k_i}{s}]G(s)}{1 + [k_p + \frac{k_i}{s}]G(s)}$$
$$= \frac{k_i + k_p s}{s^3 + 10s^2 + (20 + k_p)s + k_i} \tag{7.127}$$

The characteristic equation is

$$s^3 + 10s^2 + (20 + k_p)s + k_i = 0 \qquad (7.128)$$

The Routh array is as follows

$$
\begin{array}{ccc}
s^3 & 1 & 20 + k_p \\
s^2 & 10 & k_i \\
s^1 & 20 + k_p - \frac{k_i}{10} & 0 \\
s^0 & k_i &
\end{array}
\qquad (7.129)
$$

Applying the Routh-Hurwitz conditions of absolute stability,

(a) All coefficients in the characteristic equations should be positive, *i.e.,*

$$20 + k_p > 0, \quad k_i > 0$$

(b) All elements in the first column of the Routh array should be positive, *i.e.,*

$$20 + k_p - \frac{k_i}{10} > 0, \quad k_i > 0$$

Thus, we can write the conditions of the system stability as

$$
\begin{aligned}
k_p &> \frac{k_i}{10} - 20 \\
k_i &> 0
\end{aligned}
\qquad (7.130)
$$

The final value of X for a unit step input is

$$
\begin{aligned}
X(\infty) &= \lim_{s \to 0} sM(s)F(s) = \lim_{s \to 0} s \left[\frac{k_i + k_p s}{s^3 + 10s^2 + (20 + k_p)s + k_i} \right] \frac{1}{s} \\
&= \frac{k_i}{k_i} = 1 \qquad (7.131)
\end{aligned}
$$

The steady state error e_{ss} is

$$e_{ss} = 1 - X(\infty) = 1 - 1 = 0 \qquad (7.132)$$

For the given gains k_p=30 and k_i=70, the characteristic equation is

$$s^3 + 10s^2 + 50s + 70 = 0 \qquad (7.133)$$

Using Matlab function "roots", we find the characteristic roots as

$$s_{1,2} = -3.954 \pm 4.222, \; s_3 = -2.092 \qquad (7.134)$$

As all the roots are in the left half s-plane, the closed-loop system would be stable.

7.9 (a) The closed-loop transfer function is

$$M(s) = \frac{G(s)}{1 + G(s)} = \frac{4500K}{s^2 + 361.2s + 4500K}$$

The characteristic equation is

$$s^2 + 361.2s + 4500K = 0$$

The preceding equation is compared with the characteristic equation of a standard second-order system:

$$s^2 + 2\zeta\omega_n s + \omega_n^2 = 0$$

where ω_n is the undamped natural frequency and ζ is the damping ratio.

Thus, we can write

$$2\zeta\omega_n = 361.2$$
$$\omega_n^2 = 4500K$$

Further simplifying,

$$\omega_n = \sqrt{4500K} = 64.082\sqrt{K} \text{ rad/s}$$
$$\zeta = \frac{361.2}{2\omega_n} = \frac{361.2}{2 \times 64.082\sqrt{K}} = \frac{2.692}{\sqrt{K}}$$

(b)

The characteristic equation of the closed-loop second-order system is obtained as

$$s^2 + 361.2s + 4500K = 0$$

The Routh array is

$$
\begin{array}{lcc}
s^2 & 1 & 4500K \\
s^1 & 361.2 & 0 \\
s^0 & 4500K &
\end{array}
$$

Applying the Routh-Hurwitz criterion of absolute stability,

All elements in the first column of the Routh array should be positive, *i.e.,*

$$4500K > 0 \Rightarrow K > 0$$

Thus, we can write the conditions of the system stability as

$$K \in (0, \infty)$$

The characteristic equation of the closed-loop second-order system is obtained as

$$s^3 + 3408.3s^2 + 1,204,000s + 1.5 \times 10^7 K = 0$$

The Routh array is

$$
\begin{array}{lcc}
s^3 & 1 & 1,204,000 \\
s^2 & 3408.3 & 1.5 \times 10^7 \\
s^1 & 1,204,000 - \frac{1.5 \times 10^7}{3408.3} & 0 \\
s^0 & 1.5 \times 10^7 K &
\end{array}
$$

Applying the Routh-Hurwitz criterion of absolute stability,

All elements in the first column of the Routh array should be positive, *i.e.,*

$$1,204,000 - \frac{1.5 \times 10^7}{3408.3} > 0 \Rightarrow 273.573 > K; \ 1.5 \times 10^7 K > 0 \Rightarrow K > 0$$

Thus, we can write the conditions of the system stability as

$$0 < K < 273.573$$

Thus, the second-order approximation is only valid for K range $0 < K < 273.573$.

(c)

The dominant roots are $s_1 = 186.53 + j192$ and $s_2 = 186.53 - j192$. The corresponding equivalent damping ratio is 0.694.

7.10 The equations of motion for the system are obtained as follows:

System: β_1

$$\beta_1'' + M_{p1}\left(\frac{L_2}{L_1}\right)\cos(\beta_1 - \beta_2)\beta_2'' + M_{p1}\left(\frac{L_2}{L_1}\right)\left\{\beta_2'(2 + \beta_2')sin(\beta_1 - \beta_2)\right.$$
$$\left. + 3sin\beta_1cos\beta_2\right\} + 3sin\beta_1cos\beta_1 = 0$$
$$(7.135)$$

System: β_2

$$\beta_2'' + M_{p2}\left(\frac{L_1}{L_2}\right)\cos(\beta_2 - \beta_1)\beta_1'' + M_{p2}\left(\frac{L_1}{L_2}\right)\left\{\beta_1'(2 + \beta_1')sin(\beta_2 - \beta_1)\right.$$
$$\left. + 3sin\beta_2cos\beta_1\right\} + 3sin\beta_2cos\beta_2 = 0$$
$$(7.136)$$

where

$$M_{p1} = \frac{m_3}{m_2 + m_3}, \quad M_{p2} = \frac{m_1}{m_1 + m_2} \qquad (7.137)$$

and $()' = d()/d\theta; ()'' = d^2()/d\theta^2$. θ is the angle with respect to a reference line.

Solving β_1'' and β_2'' using Eqs. (7.135) and (7.136), we obtain

$$[1 - M_{p1}M_{p2}cos^2(\beta_1 - \beta_2)]\beta_1'' - M_{p1}M_{p2}cos(\beta_1 - \beta_2)\Big\{\beta_1'(2 + \beta_1')sin(\beta_2 - \beta_1)$$

$$+ 3sin\beta_2cos\beta_1\Big\} + M_{p1}\left(\frac{L_2}{L_1}\right)\Big\{\beta_2'(2 + \beta_2')sin(\beta_1 - \beta_2) + 3sin\beta_1cos\beta_2\Big\}$$

$$- 3M_{p1}\left(\frac{L_2}{L_1}\right)cos(\beta_1 - \beta_2)sin\beta_2cos\beta_2 + 3sin\beta_1cos\beta_1 = 0 \quad (7.138)$$

$$[1 - M_{p1}M_{p2}cos^2(\beta_2 - \beta_1)]\beta_2'' - M_{p1}M_{p2}cos(\beta_2 - \beta_1)\Big\{\beta_2'(2 + \beta_2')sin(\beta_1 - \beta_2)$$

$$+ 3sin\beta_1cos\beta_2\Big\} + M_{p2}\left(\frac{L_1}{L_2}\right)\Big\{\beta_1'(2 + \beta_1')sin(\beta_2 - \beta_1) + 3sin\beta_2cos\beta_1\Big\}$$

$$- 3M_{p2}\left(\frac{L_1}{L_2}\right)cos(\beta_2 - \beta_1)sin\beta_1cos\beta_1 + 3sin\beta_2cos\beta_2 = 0 \quad (7.139)$$

To design a linear controller for the system, we linearize the preceding equations with respect to null reference state and we have the resulting linear equations with control inputs u_1 and u_2 as

$$[1 - M_{p1}M_{p2}]\delta\beta_1'' - 3M_{p1}[M_{p2} + \left(\frac{L_2}{L_1}\right)]\delta\beta_2 + 3[1 + M_{p1}\left(\frac{L_2}{L_1}\right)]\delta\beta_1 = u_1$$
$$(7.140)$$

$$[1 - M_{p1}M_{p2}]\delta\beta_2'' - 3M_{p2}[M_{p1} + \left(\frac{L_1}{L_2}\right)]\delta\beta_1 + 3[1 + M_{p2}\left(\frac{L_1}{L_2}\right)]\delta\beta_2 = u_2$$
$$(7.141)$$

Comparing these equations with the desired performance specified by the second order linear system with the given closed-loop damping ratio and frequency:

$$\delta\beta_1'' + 2\zeta_1\omega_1\delta\beta_1' + \omega_1^2\delta\beta_1 = 0 \qquad (7.142)$$
$$\delta\beta_2'' + 2\zeta_2\omega_2\delta\beta_2' + \omega_2^2\delta\beta_2 = 0 \qquad (7.143)$$

we obtain the following control laws

$$u_1 = - [1 - M_{p1}M_{p2}][2\zeta_1\omega_1\delta\beta_1' + \omega_1^2\delta\beta_1] - 3M_{p1}[M_{p2} + \left(\frac{L_2}{L_1}\right)]\delta\beta_2$$

$$+ 3[1 + M_{p1}\left(\frac{L_2}{L_1}\right)]\delta\beta_1 \qquad (7.144)$$

$$u_2 = - [1 - M_{p1}M_{p2}][2\zeta_2\omega_2\delta\beta_2' + \omega_1^2\delta\beta_2] - 3M_{p2}[M_{p1} + \left(\frac{L_1}{L_2}\right)]\delta\beta_1$$

$$+ 3[1 + M_{p2}\left(\frac{L_1}{L_2}\right)]\delta\beta_2 \qquad (7.145)$$

To design PD controllers to stabilize the system along local vertical ($\beta_1 = \beta_2 = 0$) within $\pm 0.01^\circ$ in half an orbit, we specify the desired damping ratio and frequency as follows:

Maximum overshoot: $M_p = 0.01 \times \pi/180 = 1.7453 \times 10^{-4}$ rad.

Settling time (θ): $t_{ds} = \pi$ rad.

$$\zeta \geq \sqrt{\frac{(lnM_p/\pi)^2}{1 + (lnM_p/\pi)^2}} = 0.9948 \qquad (7.146)$$

$$t_s = \frac{4.6}{\zeta\omega} \leq t_{ds} \quad 1\% \text{ settling time}$$

$$\frac{4.6}{0.9948\pi} \leq \omega$$

$$\Rightarrow \omega \geq 1.4719 \qquad (7.147)$$

Thus,

$$\omega_1 = \omega_2 = 1.5, \zeta_1 = \zeta_2 = 1 \qquad (7.148)$$

Next we apply the control laws in the nonlinear system defined by Eqs. (7.138) and (7.139) as follows:

$$[1 - M_{p1}M_{p2}cos^2(\beta_1 - \beta_2)]\beta_1'' - M_{p1}M_{p2}cos(\beta_1 - \beta_2)\Big\{\beta_1'(2 + \beta_1')sin(\beta_2 - \beta_1)$$

$$+ 3sin\beta_2cos\beta_1\Big\} + M_{p1}\left(\frac{L_2}{L_1}\right)\Big\{\beta_2'(2 + \beta_2')sin(\beta_1 - \beta_2) + 3sin\beta_1cos\beta_2\Big\}$$

$$- 3M_{p1}\left(\frac{L_2}{L_1}\right)cos(\beta_1 - \beta_2)sin\beta_2cos\beta_2 + 3sin\beta_1cos\beta_1 = u_1 \quad (7.149)$$

$$[1 - M_{p1}M_{p2}cos^2(\beta_2 - \beta_1)]\beta_2'' - M_{p1}M_{p2}cos(\beta_2 - \beta_1)\Big\{\beta_2'(2 + \beta_2')sin(\beta_1 - \beta_2)$$

$$+ 3sin\beta_1cos\beta_2\Big\} + M_{p2}\left(\frac{L_1}{L_2}\right)\Big\{\beta_1'(2 + \beta_1')sin(\beta_2 - \beta_1) + 3sin\beta_2cos\beta_1\Big\}$$

$$- 3M_{p2}\left(\frac{L_1}{L_2}\right)cos(\beta_2 - \beta_1)sin\beta_1cos\beta_1 + 3sin\beta_2cos\beta_1 = u_2 \quad (7.150)$$

The following date has been considered for system masses:

$m_1 \gg m_2, m_3$ and $m_2 = m_3$ leading to $M_{p1} = 0.5$ and $M_{p2} = 1$.

(a) linear equations of motion, 5° initial attitude error in β_1, no disturbance

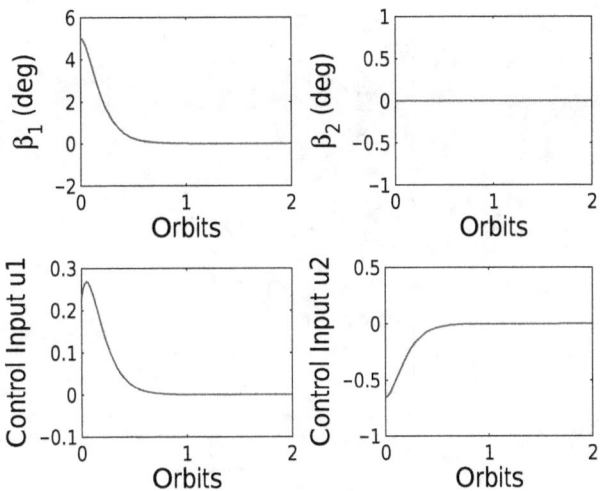

Figure 7.1: System Response

(b) linear equations of motion, 5° initial attitude error in β_2, no disturbance

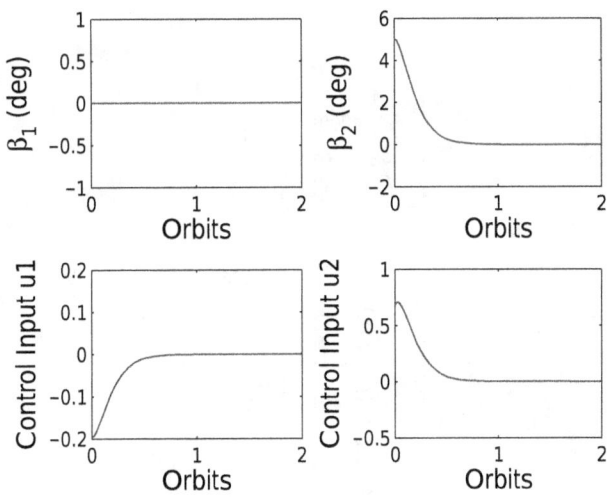

Figure 7.2: System Response

(c) nonlinear equations of motion, 5° initial attitude errors in β_1 and β_2

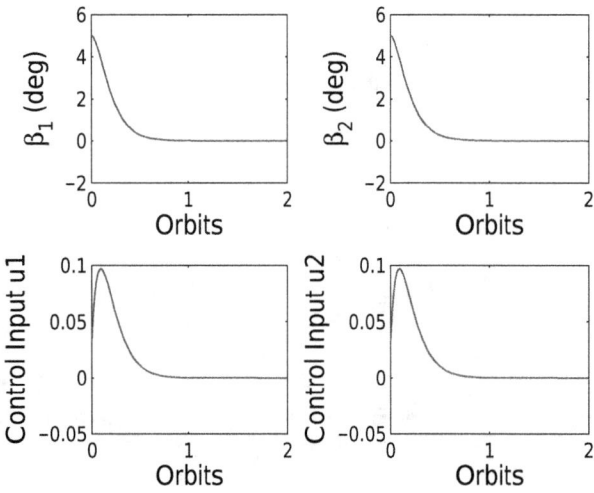

Figure 7.3: System Response

(d) nonlinear equations of motion, 135° initial attitude errors in β_1 and β_2, 1 Nm disturbance torques along β_1 and β_2

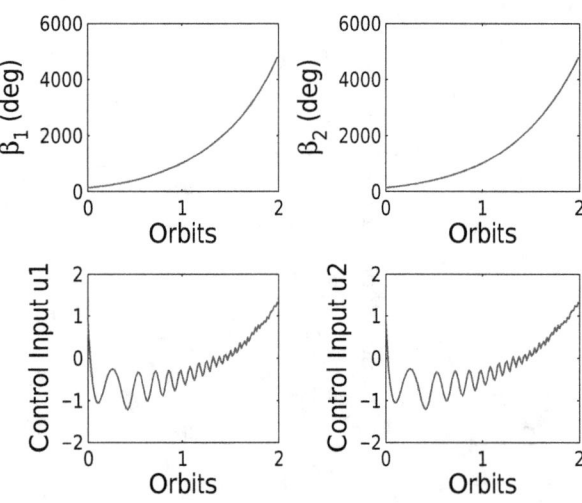

Figure 7.4: System Response

7.12 The pitch equation motion of the spacecraft is

$$\delta\alpha'' - 3\frac{k_2 - k_1}{1 - k_1 k_2}\delta\alpha = u_\alpha = -k_{p_\alpha}\delta\alpha - k_{d_\alpha}\delta\alpha' \qquad (7.151)$$

The closed-loop pitch characteristic equation is derived as

$$s^2 + k_{d_\alpha}s + 3\left(\frac{k_2 - k_1}{1 - k_1 k_2} + k_{p_\alpha}\right) = 0 \qquad (7.152)$$

Based on the preceding characteristic equation, the controller gains for the stable response of the system should be

$$k_{d_\alpha} > 0 \qquad (7.153)$$

$$k_{p_\alpha} > \frac{k_1 - k_2}{1 - k_1 k_2} \qquad (7.154)$$

The roll and yaw equations motion of the spacecraft are

$$\delta\phi'' + (1 - k_1)\delta\gamma' + 4k_1\delta\phi = u_\phi = -k_{p_\phi}\delta\phi - k_{d_\phi}\delta\phi'$$

$$\delta\gamma'' + (k_3 - 1)\delta\phi' + k_3\delta\gamma = u_\gamma = -k_{p_\gamma}\delta\gamma - k_{d_\gamma}\delta\gamma' \qquad (7.155)$$

We write the Eqs.(7.155) in the form

$$X' = AX \qquad (7.156)$$

where X is a state vector given by

$$X = \begin{Bmatrix} \delta\phi \\ \delta\phi' \\ \delta\gamma \\ \delta\gamma' \end{Bmatrix} \qquad (7.157)$$

and matrix A is

$$A = \begin{bmatrix} 0 & 1 & 0 & 0 \\ -4k_1 - k_{p_\phi} & -k_{d_\phi} & 0 & k_1 - 1 \\ 0 & 0 & 0 & 1 \\ 0 & 1 - k_3 & -k_3 - k_{p_\gamma} & -k_{d_\gamma} \end{bmatrix} \qquad (7.158)$$

To obtain the eigenvalues, we write

$$|A - sI| = \begin{vmatrix} -s & 1 & 0 & 0 \\ -4k_1 & -s & 0 & k_1 - 1 \\ 0 & 0 & -s & 1 \\ 0 & 1 - k_2 & -k_2 & -s \end{vmatrix} = 0 \qquad (7.159)$$

The closed-loop characteristic equation we obtain is

$$s^4 + (k_{d_\phi} + k_{d_\phi})s^3 + (1 + 3k_1 + k_1 k_3 + -k_{p_\phi} + k_{p_\gamma} + k_{d_\phi} k_{d_\gamma})s^2$$
$$+ [k_{d_\phi}(k_3 + k_{p_\gamma}) + k_{d_\gamma}(4k_1 + k_{p_\phi})]s + (4k_1 + k_{p_\phi})(k_3 + k_{p_\gamma}) = 0$$
$$(7.160)$$

Chapter 8

Formation Flying

Problem Set 8

8.1 (I) For the given system

$$\mathbf{A} = \begin{bmatrix} 0 & 1 & 0 & 0 \\ 3\dot{\theta}^2 & 0 & 0 & 2\dot{\theta} \\ 0 & 0 & 0 & 1 \\ 0 & -2\dot{\theta} & k_p & 0 \end{bmatrix}$$

The corresponding characteristic equation of the given system is

$$s^4 + (\dot{\theta}^2 - k_p)s^2 + 3\dot{\theta}^2 k_p = 0 \qquad (8.1)$$

The Routh array is as follows

$$
\begin{array}{cccc}
s^4 & 1 & \dot{\theta}^2 - k_p & 3\dot{\theta}^2 k_p \\
s^3 & 0 & 0 & 0
\end{array}
$$

We can not proceed further as all elements in the s^3-row are zero. So, we construct auxiliary function $A_1(s)$ using the elements in the s^4-row as

$$A_1(s) = s^4 + (\dot{\theta}^2 - k_p)s^2 + 3\dot{\theta}^2 k_p \qquad (8.2)$$

Next, we differentiate the preceding equation with respect to s

$$\frac{dA(s)}{ds} = 4s^3 + 2(\dot{\theta}^2 - k_p)s \qquad (8.3)$$

The elements of the s^3-row of the above Routh array are taken from the coefficients in Eq. (8.3) and thus, the Routh array is

$$
\begin{array}{cccc}
s^4 & 1 & \dot{\theta}^2 - k_p & 3\dot{\theta}^2 k_p \\[2mm]
s^3 & 4 & 2(\dot{\theta}^2 - k_p) & 0 \\[2mm]
\dfrac{\dot{\theta}^2 - k_p}{2} & 3\dot{\theta}^2 k_p & 0 & \\[2mm]
s^1 & \dfrac{2[(\dot{\theta}^2 - k_p)^2 - 12\dot{\theta}^2 k_p]}{\dot{\theta}^2 - k_p} & 0 & \\[2mm]
s^0 & 3\dot{\theta}^2 k_p & &
\end{array}
\qquad (8.4)
$$

Applying the Routh-Hurwitz criterion of absolute stability,

(a) All coefficients in the characteristic equations should be positive, *i.e.*,
$$\dot{\theta}^2 - k_p > 0 \Rightarrow k_p < \dot{\theta}^2; \ 3\dot{\theta}^2 k_p > 0 \Rightarrow k_p > 0$$

(b) All elements in the first column of the Routh array should be positive, *i.e.*,
$$\dot{\theta}^2 - k_p > 0 \Rightarrow k_p < \dot{\theta}^2,$$

and
$$(\dot{\theta}^2 - k_p)^2 - 12\dot{\theta}^2 k_p > 0 \Rightarrow k_p^2 - 14\dot{\theta}^2 k_p + \dot{\theta}^2 > 0$$

$$\Rightarrow (k_p - 13.928\dot{\theta}^2)(k_p - 0.0718\dot{\theta}^2) > 0$$

$$\Rightarrow k_p > 13.928\dot{\theta}^2 \ \text{ or } \ k_p < 0.0718\dot{\theta}^2$$

Thus, we can write the conditions of the system stability as

$$0 < k_p < 0.0718\dot{\theta}^2 \qquad (8.5)$$

(II) The characteristic equation of the given system is obtained as

$$s^4 - k_d s^3 + (\dot{\theta}^2 - k_p)s^2 + (3\dot{\theta}^2 k_d)s + 3\dot{\theta}^2 k_p = 0 \qquad (8.6)$$

The Routh array is

$$
\begin{array}{llll}
s^4 & 1 & \dot{\theta}^2 - k_p & 3\dot{\theta}^2 k_p \\[4pt]
s^3 & -k_d & 3\dot{\theta}^2 k_d & 0 \\[4pt]
s^2 & 4\dot{\theta}^2 - k_p & 3\dot{\theta}^2 k_p & 0 \\[4pt]
s^1 & \dfrac{12\dot{\theta}^4 k_d}{4\dot{\theta}^2 - k_p} & 0 & 0 \\[10pt]
s^0 & 3\dot{\theta}^2 k_p & &
\end{array}
\tag{8.7}
$$

Applying the Routh-Hurwitz criterion of absolute stability,

(a) All coefficients in the characteristic equations should be positive, *i.e.,*

$$
-k_d > 0 \Rightarrow -k_d > 0; \theta^2 - k_p > 0 \Rightarrow k_p < \dot{\theta}^2; (3\dot{\theta}^2 k_d) > 0 \Rightarrow k_d > 0
$$

and
$$
3\dot{\theta}^2 k_p > 0 \Rightarrow k_p > 0
$$

(b) All elements in the first column of the Routh array should be positive, *i.e.,*

$$
-k_d > 0 \Rightarrow k_d < 0; 4\dot{\theta}^2 - k_p > 0 \Rightarrow k_p < 4\dot{\theta}^2; 12\dot{\theta}^4 k_d > 0 \Rightarrow k_d > 0
$$

and
$$
3\dot{\theta}^2 k_p > 0 \Rightarrow k_p > 0
$$

Thus, we can write the conditions of the system stability as

$$
0 < k_p < \dot{\theta}^2; k_d < 0, k_d > 0
\tag{8.8}
$$

Thus, the system will be unstable for any values of k_d.

(III) The characteristic equation of the given system is obtained as

$$
s^4 - k_{d2}s^3 + (\dot{\theta}^2 - k_{p2} - 2\dot{\theta}k_{d1})s^2 + (3\dot{\theta}^2 k_{d2} - 2\dot{\theta}k_{p1})s + 3\dot{\theta}^2 k_{p2} = 0
\tag{8.9}
$$

Taking $k_{d3} = -k_{d2}$, the Routh array is

$$
\begin{array}{cccc}
s^4 & 1 & \dot{\theta}^2 + k_{p2} - 2\dot{\theta}k_{d1} & 3\dot{\theta}^2 k_{p2} \\
s^3 & k_{d3} & -3\dot{\theta}^2 k_{d3} - 2\dot{\theta}k_{p1} & 0 \\
s^3 & 4\dot{\theta}^2 - k_{p2} - 2\dot{\theta}k_{d1} & 3\dot{\theta}^2 k_{p2} & 0 \\
    & \quad + \dfrac{2\dot{\theta}k_{p1}}{k_{d3}} & & \\[2em]
s^1 & -3\dot{\theta}^2 k_{d3} - 2\dot{\theta}k_{p1} & 0 & 0 \\
    & \quad -\dfrac{3\dot{\theta}k_{p2}k_{d3}}{4\dot{\theta}^2 - k_{p2} - 2\dot{\theta}k_{d1} + 2\dot{\theta}k_{p1}/k_{d3}} & & \\[2em]
s^0 & 3\dot{\theta}^2 k_{p2} & &
\end{array}
$$

$$(8.10)$$

Applying the Routh-Hurwitz criterion of absolute stability,

(a) All coefficients in the characteristic equations should be positive, *i.e.,*

$$k_{d3} > 0;$$

$$\dot{\theta}^2 - k_{p2} - 2\dot{\theta}k_{d1} > 0 \Rightarrow k_{d1} < \frac{\dot{\theta}^2 - k_{p2}}{2\dot{\theta}};$$

$$-3\dot{\theta}^2 k_{d3} - 2\dot{\theta}k_{p1} > 0 \Rightarrow k_{p1} < \frac{-3\dot{\theta}^2 k_{d3}}{2\dot{\theta}}$$

and

$$3\dot{\theta}^2 k_{p2} > 0 \Rightarrow k_{p2} > 0$$

(b) All elements in the first column of the Routh array should be positive, *i.e.,*

$$k_{d3} > 0;$$

$$4\dot{\theta}^2 - k_{p2} - 2\dot{\theta}k_{d1} + \frac{2\dot{\theta}k_{p1}}{k_{d3}} > 0 \Rightarrow k_{p1} > \frac{k_{d3}}{2\dot{\theta}}\left(-4\dot{\theta}^2 + k_{p2} + 2\dot{\theta}k_{d1}\right);$$

$$-3\dot{\theta}^2 k_{d3} - 2\dot{\theta}k_{p1} - \frac{3\dot{\theta}k_{p2}k_{d3}}{4\dot{\theta}^2 - k_{p2} - 2\dot{\theta}k_{d1} + \dfrac{2\dot{\theta}k_{p1}}{k_{d3}}} > 0$$

$$\Rightarrow k_{d1} < \frac{1}{\dot{\theta}} \left(4\dot{\theta}^2 - k_{p2} + \frac{2\dot{\theta}k_{p1}}{k_{d3}} + \frac{3\dot{\theta}^2 k_{p2} k_{d3}}{3\dot{\theta}^2 k_{d3} + 2\dot{\theta}k_{p1}} \right)$$

and

$$3\dot{\theta}^2 k_{p2} > 0 \Rightarrow k_{p2} > 0$$

Thus, we can write the conditions of the system stability as

$$\frac{-3\dot{\theta}^2 k_{d3}}{2\dot{\theta}} > k_{p1} > \frac{k_{d3}}{2\dot{\theta}} \left(-4\dot{\theta}^2 + k_{p2} + 2\dot{\theta}k_{d1} \right)$$

$$\Rightarrow \frac{-3\dot{\theta}^2 k_{d3}}{2\dot{\theta}} > \frac{k_{d3}}{2\dot{\theta}} \left(-4\dot{\theta}^2 + k_{p2} + 2\dot{\theta}k_{d1} \right)$$

$$\Rightarrow k_{p2} > \dot{\theta}^2 - 2\dot{\theta}k_{d1}$$

We also have

$$0 < k_{p2} < \dot{\theta}^2 - 2\dot{\theta}k_{d1}$$

$$k_{d1} < \frac{1}{ds\dot{\theta}} \left(4\dot{\theta}^2 - k_{p2} + \frac{2\dot{\theta}k_{p1}}{k_{d3}} + \frac{3\dot{\theta}^2 k_{p2} k_{d3}}{3\dot{\theta}^2 k_{d3} + 2\dot{\theta}k_{p1}} \right) \text{ and } k_{d1} < \frac{\dot{\theta}^2 - k_{p2}}{2\dot{\theta}}$$

Thus, the system will be unstable for any values of k_d.

Index